142
Advances in Biochemical Engineering/Biotechnology

Aims and Scope

This book series reviews current trends in modern biotechnology and biochemical engineering. Its aim is to cover all aspects of these interdisciplinary disciplines, where knowledge, methods and expertise are required from chemistry, biochemistry, microbiology, molecular biology, chemical engineering and computer science.

Volumes are organized topically and provide a comprehensive discussion of developments in the field over the past 3–5 years. The series also discusses new discoveries and applications. Special volumes are dedicated to selected topics which focus on new biotechnological products and new processes for their synthesis and purification.

In general, volumes are edited by well-known guest editors. The series editor and publisher will, however, always be pleased to receive suggestions and supplementary information. Manuscripts are accepted in English.

In references, Advances in Biochemical Engineering/Biotechnology is abbreviated as *Adv. Biochem. Engin./Biotechnol.* and cited as a journal.

More information about this series at http://www.springer.com/series/10

Axel Schippers · Franz Glombitza
Wolfgang Sand
Editors

Geobiotechnology II

Energy Resources, Subsurface Technologies, Organic Pollutants and Mining Legal Principles

With contributions by

Petra Bombach · Alexander Galushko · Claudia Gniese
Nils Hoth · Martin Krüger · Jana Rakoczy
Hans-Hermann Richnow · Giovanni Rossi
Michael Schlömann · Reinhard Schmidt
Michael Siegert · Jana Sitte · Carsten Vogt

Springer

Editors
Axel Schippers
BGR—Fachbereich Geochemie d. Rohstoffe
Hannover
Germany

Wolfgang Sand
Fakultät für Chemie, Biofilm Centre
Universität Duisburg-Essen
Essen
Germany

Franz Glombitza
G.E.O.S. Ingenieurgesellschaft mbH
Halsbrücke
Germany

ISSN 0724-6145 ISSN 1616-8542 (electronic)
ISBN 978-3-662-44473-3 ISBN 978-3-662-44474-0 (eBook)
DOI 10.1007/978-3-662-44474-0

Library of Congress Control Number: 2014943414

Springer Heidelberg New York Dordrecht London

Printed on acid-free paper

Springer is part of Springer Science+Business Media (www.springer.com)

Preface

Mineral and energy resources are increasingly being exploited to meet the demands of a worldwide growing population and economy. Despite technological developments, these raw materials cannot, or can only partly, be substituted by renewable resources within the next few decades. Thus, the efficient recovery and processing of mineral and energy resources, as well as recycling, are nowadays of significant importance in many countries.

Geobiotechnology can significantly contribute to new developments in this field and can be described as biotechnology in the geological context. This technology mainly takes advantage of the biological activity relevant for geochemical processes. Microorganisms control natural biogeochemical cycles and by doing so they contribute to the formation and alteration of metal, oil, coal, and phosphor deposits. Geobiotechnology comprises microbial processes in these deposits as well as in mining and environment. The interactions of microorganisms with raw materials enable an efficient geobiotechnological recovery of metals, oil and gas.

The five chapters of this volume describe and summarize the scientific background and recent developments in metal bioleaching, bioextraction, biomineralization and bioremediation as well as in microbial enhanced oil and gas recovery (MEOR). Microbial processes in the underground and deposits, potentially used for the storage of raw materials or residues, or use of geothermal energy are also covered, including a chapter about basic mining legal principles.

The idea for this volume originated from the temporary working group Geobiotechnology of the German organisation DECHEMA e.V. Since many authors of this volume are active in this working group, geobiotechnological processes and applications are often described using examples from Germany and Europe.

The chapter on coal biotechnology is authored by the late Giovanni Rossi. He died in summer 2013 and could not live to see the publication. Giovanni Rossi was a dear friend, esteemed colleague, consummate engineer and researcher, and a pioneer in the field of biohydrometallurgy. We feel honored that he was able to finalize his contribution to this book. We dedicate this book in memory of Giovanni Rossi.

Axel Schippers
Franz Glombitza
Wolfgang Sand

Contents

Starting Up Microbial Enhanced Oil Recovery 1
Michael Siegert, Jana Sitte, Alexander Galushko and Martin Krüger

Relevance of Deep-Subsurface Microbiology for Underground Gas Storage and Geothermal Energy Production 95
Claudia Gniese, Petra Bombach, Jana Rakoczy, Nils Hoth, Michael Schlömann, Hans-Hermann Richnow and Martin Krüger

Bioremediation via in situ Microbial Degradation of Organic Pollutants 123
Carsten Vogt and Hans Hermann Richnow

The Microbial Desulfurization of Coal 147
Giovanni Rossi

Description of Basic Mining Legal Principles 169
Reinhard Schmidt

Index 197

Adv Biochem Eng Biotechnol (2014) 142: 1–94
DOI: 10.1007/10_2013_256

Published Online: 16 November 2013

Starting Up Microbial Enhanced Oil Recovery

Michael Siegert, Jana Sitte, Alexander Galushko and Martin Krüger

Abstract This chapter gives the reader a practical introduction into microbial enhanced oil recovery (MEOR) including the microbial production of natural gas from oil. Decision makers who consider the use of one of these technologies are provided with the required scientific background as well as with practical advice for upgrading an existing laboratory in order to conduct microbiological experiments. We believe that the conversion of residual oil into natural gas (methane) and the in situ production of biosurfactants are the most promising approaches for MEOR and therefore focus on these topics. Moreover, we give an introduction to the microbiology of oilfields and demonstrate that in situ microorganisms as well as injected cultures can help displace unrecoverable oil in place (OIP). After an initial research phase, the enhanced oil recovery (EOR) manager must decide whether MEOR would be economical. MEOR generally improves oil production but the increment may not justify the investment. Therefore, we provide a brief economical assessment at the end of this chapter. We describe the necessary state-of-the-art scientific equipment to guide EOR managers towards an appropriate MEOR strategy. Because it is inevitable to characterize the microbial community of an oilfield that should be treated using MEOR techniques, we describe three complementary start-up approaches. These are: (i) culturing methods, (ii) the characterization of microbial communities and possible bio-geochemical pathways by using molecular biology methods, and (iii) interfacial tension measurements. In conclusion, we hope that this chapter will facilitate a decision on whether to launch MEOR activities. We also provide an update on relevant literature for experienced

M. Siegert (✉)
Department of Civil and Environmental Engineering,
The Pennsylvania State University, 127 Sackett Building,
University Park, PA 16802, USA
e-mail: michael@siegert.org

J. Sitte · M. Krüger
Bundesanstalt für Geowissenschaften und Rohstoffe,
Stilleweg 2, 30655 Hannover, Germany
e-mail: martin.krueger@bgr.de

A. Galushko
Department of Microbial Ecology, Faculty of Life Sciences,
University of Vienna, Althanstr. 14, 1090 Vienna, Austria
e-mail: galushko@microbial-ecology.net

MEOR researchers and oilfield operators. Microbiologists will learn about basic principles of interface physics needed to study the impact of microorganisms living on oil droplets. Last but not least, students and technicians trying to understand processes in oilfields and the techniques to examine them will, we hope, find a valuable source of information in this review.

Keywords Anaerobic cultivation • Interfacial tension • Methanogenic hydrocarbon degradation • Microbial enhanced oil recovery • Peak oil

Abbreviations

16S rRNA	Ribosomal RNA of a sedimentation rate of 16 Svedberg
A	Surface area of an oil droplet
alk	Alkane hydroxylase gene
apsA	Adenosine-5′-phosphosulfate (APS) reductase gene
bbl	Barrel (oil)
CARD-FISH	Catalyzed reporter deposition, fluorescence in situ hybridization
cDNA	Complementary DNA for an RNA strand
CMC	Critical micelle concentration
dsrAB	Dissimilatory (bi)sulfite reductase gene
DGGE	Denaturing gradient gel electrophoresis
DNA	Deoxyribonucleic acid
E	Elasticity
EDTA	Ethylenediaminetetraacetate
EOR	Enhanced oil recovery
EPS	Extracellular polymeric substances
f	Frequency
FISH	Fluorescence in situ hybridization
g	Gravity force
ΔG	Gibbs free energy
γ	Interfacial tension
HOT	Hot oil treatment
licA	Lichenysin A synthetase gene
μ	Viscosity
mcr	Methyl coenzyme M reductase gene
MEOR	Microbial enhanced oil recovery
MIC	Microbial influenced corrosion
MPN	Most probable number
mRNA	Messenger RNA
nar	Nitrate reductase gene
nir	Nitrite reductase gene
nor	Nitrite oxidoreductase gene
nos	Nitric oxide synthase gene
nrf	Nitrite reductase gene (to ammonium)
NTA	Nitrilotriacetic acid

OIP	Oil in place
omc	Outer-membrane cytochrome gene
omp	Outer-membrane multicopper protein gene
P	Pressure
PCR	Polymerase chain reaction
PLFA	Phospholipid-derived fatty acids
psia	Pounds per square inch
qPCR	Quantitative PCR
RNA	Ribonucleic acid
RT-qPCR	Reverse transcription-qPCR
SAC	Surface active compound
sfp	Surfactin synthesizing protein gene
SIP	Stable isotope probing
srf	*sfp* operon (*sfp* gene cluster)
θ	Contact angle
SSU	Small subunit
T-RFLP	Terminal restriction fragment length polymorphism
V	Volume

Contents

1 Introduction ... 3
2 Microbial Life in Oilfields ... 5
2.1 Environmental Conditions in Oil Reservoirs ... 6
2.2 Oil Reservoir Microorganisms ... 8
2.3 Microbiological Sampling Techniques for Oil Reservoirs ... 9
2.4 Characterization of Microbial Populations in Oilfields ... 10
3 Selecting an Appropriate MEOR Approach ... 22
3.1 Conversion of Oil or Coal into Natural Gas (Methane) ... 22
3.2 Oil Swelling by Microbial Production of Gases ... 24
3.3 Detaching Residual Hydrocarbons from the Mineral Matrix (Capillary Number N_{CA}) ... 25
3.4 Effects of Oil and Water Viscosity ... 34
3.5 Porespace Alteration by Plugging or Rock Dissolution ... 36
4 The Economy of MEOR ... 37
5 Conclusions ... 38
6 Appendix ... 69
References ... 69

1 Introduction

Expanding worldwide wealth will result in an increased global energy demand of 49 % by 2035 compared with 2007 [1]. The worldwide need for energy from liquid hydrocarbons will likely increase from 1.9×10^{11} GJ (2010) to 2.3×10^{11} GJ y^{-1} (2035), from coal it will incline from 1.4×10^{11} to 2.2×10^{11} GJ y^{-1}, and from natural gas (methane) from 1.4×10^{11} to 1.7×10^{11} GJ y^{-1}. At the same time, new resources of fossil oil become more and more limited. The resulting limit of oil supply is commonly known as peak oil, which was first predicted in the 1920s. Despite impressive technological progress in oilfield exploration resulting in the steady incline of fossil oil production (Fig. 1) the incremental production will not keep pace with demand.

At the same time, up to two thirds of the original oil in place (OIP) in producing oilfields remains impossible to extract using current recovery technologies [3] such as enhanced oil recovery (EOR). Traditionally, this term is used for physical and chemical methods aiming to increase oil recovery from matured oilfields. They are relatively simple to apply (e.g., CO_2 injection) but can also be expensive (e.g., costly polymers) and risky (e.g., when a polymer closes an extraction conduit or CO_2 escapes uncontrolled from a geological formation). Production and transport of EOR chemicals are the main cost factors and could be lowered if these chemicals were produced in situ, that is, at the oil droplet. Microorganisms are miniaturized biochemical machines that are able to produce the desired compounds in situ. This notion is not new [4]. Microbial enhanced oil recovery (MEOR) has been studied in field trials since the 1950s but never experienced a breakthrough as an industrial standard [5, 6]. This is not due to the lack of promising pilot studies [7] but due to the low oil price throughout the last century (Fig. 2) and the availability of inexpensive and easy-to-control CO_2-injection EOR.

The term MEOR encompasses several different approaches, including the five listed below:

(I) The production of acids that can dissolve carbonates which will liberate the entrapped oil in place (OIP).
(II) The production of gasses that can swell the OIP or reduce its viscosity, as well increase the reservoir pressure.
(III) The production of organic solvents that can alter the rock wettability, reduce the viscosity of the OIP or increase the viscosity of the water.
(IV) The production of biofilms or other biopolymers that increase the water viscosity or block exhausted canals, thereby increasing the sweep efficiency.
(V) The production of surface active compounds (SAC) that reduce the interfacial tension (γ) between oil and water.
(VI) The production of methane from oil.

Herein, we discuss all of the above-mentioned methods, but focus on oil or water viscosity and surface and interface properties as well as on methane production (points II, III, V, and VI).

MEOR may require the injection of a microbial inoculum, a nutrient broth that contains microorganisms [6, 7]. Alternatively, nutrients alone may be injected to stimulate growth of certain members of the indigenous microbial community [9]. Inoculation would usually require additional carbon sources, however, the stimulation of the indigenous microbial community allows for the production of

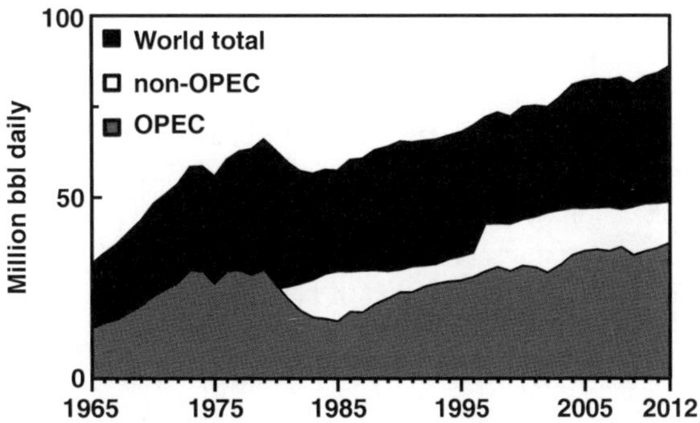

Fig. 1 Oil production worldwide during the last four decades. Data were compiled from public resources by the oil company BP (London, [2]). Non-OPEC countries include countries of the former Soviet Union. "World total" (*black*) is the sum of "OPEC" (*white*) and "non-OPEC" (*grey*)

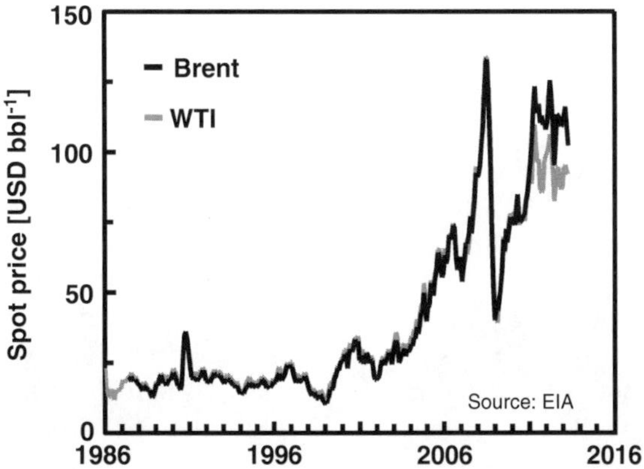

Fig. 2 Spot oil prices during the last three decades. Data were taken from the public resource of the US Energy Information Administration in June 2013 [8]. *WTI* West Texas Intermediate crude oil, *Brent* Brent crude oil

metabolites using OIP as substrate. Most reported incremental oil recovery was achieved without the need of an inoculum (15,000 bbl with inoculum versus 98,000 bbl without inoculum, [7]). Since MEOR was first proposed [4], an appreciable amount of literature has been published (for reviews see Refs. [6, 7, 10]). Although intensively studied with approximately 400 MEOR patents, only 118,000 bbl of incremental oil were reported from 400 MEOR field studies worldwide since 1970 [7] as opposed to well-established CO_2-injection EOR which recovers more than 245,000 bbl incremental oil per day, in the United States alone [11]. Although MEOR is still in its infancy, Youssef et al. [7] found that 96 % of all studied MEOR projects were successful. Obviously, MEOR can be an alternative to existing OIP extraction methods.

2 Microbial Life in Oilfields

Bastin [12] was the first to report microbial life in oilfields. Examination of microbial life in oil- and gasfields requires aseptic sampling but such sampling techniques were not developed until the 1990s. Consequently, little was known about indigenous microbial life in oilfields. The first sulfate-reducing isolates were obtained from producing wells of oilfields in the Soviet Union during the late 1980s and the early 1990s [13, 14]. At the same time, the first methanogenic archaea (methanogens) were isolated from a Californian oilfield, suggesting an indigenous origin [15]. Voordouw et al. [16] were the first to use the 16S rRNA gene cloning technique to discover sulfate reducing, sulfide oxidizing, and fermentative microorganisms in samples taken from producing wells of a Canadian oilfield. The introduction of better aseptic sampling techniques for cultivation [17] and the rapid development of molecular biology methods resulted eventually in the discovery of unexpected microbial populations [18, 19]. Moreover, some unique microbial genera such as *Geotoga* and *Petrotoga* were isolated exclusively from oil reservoirs [20]. Other closely related microorganisms were detected in distinct oilfields [21, 22]. Today it is generally accepted that microbial communities inhabit oil reservoirs and have a considerable impact on reservoir geochemistry. The occurrence of biodegraded petroleum in oilfields is the most apparent manifestation of microbial in situ activity [23]. Moreover, production of hydrogen sulfide is a consequence of microbial activity in reservoirs and can result in considerable damage to production equipment [24]. Sulfate present in injected seawater stimulates the growth of sulfate-reducing microorganisms that use hydrogen generated during steel corrosion or directly corrode steel production equipment [25], known as microbially influenced corrosion (MIC). Whether these microorganisms penetrate reservoirs or only populate production equipment has not been unequivocally clarified. Using appropriate sampling techniques, indigenous microorganisms may be distinguished from those introduced.

2.1 Environmental Conditions in Oil Reservoirs

Fossil hydrocarbons (petroleum, crude oil) originate from complex organic matter (kerogen) that formed via biological CO_2 fixation [26, 27]. Kerogen is reduced to petroleum during late diagenesis in the source rock. When petroleum escapes from the source rock, for example, due to tectonic activity, it migrates closer to the surface. Before deep drilling became economically feasible, only oil reservoirs in about 100–200 m depth could be extracted. Today, deep reservoirs are accessible as drilling depths of 8 km below the seafloor and more can be reached. The migration into layers of different geological settings is also the reason for the high diversity of oil reservoirs. Temperatures higher than 80 °C are frequently observed in oil reservoirs [23]. Such temperatures inhibit anaerobic hydrocarbon degradation in oil reservoirs [18, 23]. Biodegradation and the corresponding increase of fatty acid concentrations were not observed at reservoir temperatures higher than 80 °C [28]. Nevertheless, hyperthermophiles were found to live at 85 and 102 °C in an Alaskan oilfield [29]. However, they may not have been indigenous to the reservoir or are not restricted by the "temperature limit" for anaerobic hydrocarbon degradation.

Table 1 Gibbs free energies ($\Delta G°$) of methanogenic hydrocarbon degradation

Hydrocarbon	Reaction′	$\Delta G^{\circ\prime}$ (kJ mol^{-1})
Hexadecane	$4\,C_{16}H_{34} + 45\,H_2O \rightarrow 49\,CH_4 + 15\,HCO_3^- + 15\,H^+$	–212.84
Octadecane	$4\,C_{18}H_{38} + 51\,H_2O \rightarrow 55\,CH_4 + 17\,HCO_3^- + 17\,H^+$	–186.85
Ethylbenzene	$4\,C_8H_{10} + 33\,H_2O \rightarrow 21\,CH_4 + 11\,HCO_3^- + 11\,H^+$	–81.71
Naphthalene	$C_{10}H_8 + 12\,H_2O \rightarrow 6\,CH_4 + 4\,HCO_3^- + 4\,H^+$	–68.06
1-Methylnapthalene	$4\,C_{11}H_{10} + 51\,H_2O \rightarrow 27\,CH_4 + 17\,HCO_3^- + 17\,H^+$	–66.97
2-Methylnapthalene	$4\,C_{11}H_{10} + 51\,H_2O \rightarrow 27\,CH_4 + 17\,HCO_3^- + 17\,H^+$	–65.16
Anthracene	$4\,C_{14}H_{10} + 69\,H_2O \rightarrow 33\,CH_4 + 23\,HCO_3^- + 23\,H^+$	–64.20

Energies displayed were calculated for physiological standard conditions ($\Delta G^{\circ\prime}$, per mole hydrocarbon, pH 7, gases 1 atm, dissolved compounds 1 M, equilibrium)

Diagenesis gives all oil reservoirs a reduced redox state; that is, electron acceptors are depleted. Under such conditions, methanogenic hydrocarbon degradation prevails. Methanogenesis with hydrocarbon substrates is energetically possible but energy gains decrease when the oxidation state of the hydrocarbon substrate increases, that is, with the presence of aromatic rings (Table 1). This is reflected in the absence or strongly reduced contents of aliphatic hydrocarbons from biodegraded oilfields [30]. To prevent massive OIP degradation during MEOR, electron acceptors other than oxygen, for example, nitrate or sulfate, may be injected into reservoirs to stimulate microbial growth [31]. A general disadvantage of adding electron acceptors is the potential for accelerated corrosion of production equipment.

In deep oil reservoirs, water sometimes contains high amounts of dissolved salts ranging from 0.1 % to saturation [32]. Microorganisms are usually adapted to specific salinity and pH conditions (Table A.1). For this reason, isolates or enrichment cultures originating from the oilfield that may undergo MEOR treatment should be used.

Oil companies use computer models to calculate in situ pH including CO_2 and H_2S partial pressure. Reservoirs are characterized by a wide pH range. Many oilfield isolates tolerate a broad pH spectrum and grow very well under slightly alkaline conditions (Table A.1). In a slightly alkaline pH range biosurfactants and surfactants that are already present in the reservoir perform better. However, microorganisms that are to be injected into an oil reservoir must be able to grow under reservoir pH. Hence, the best MEOR strategy is to isolate microorganisms and culture them under reservoir conditions to demonstrate their applicability as shown in numerous laboratory studies [33–41]. Unfortunately, such studies often use aerobic strains or strains that grow anaerobically using organic matter such as sugars, yeast extract, and others. This is due to short-term research projects that only allow incubations with easily degradable substrates.

2.2 Oil Reservoir Microorganisms

Surprisingly, microorganisms isolated from distant oil reservoirs are often closely related [22, 42]. Only prokaryotes were isolated from oilfields so far and eukaryotes, such as yeasts were not reported indicating that they do not play a significant role in oilfields. Due to long isolation procedures of anaerobic hydrocarbon degrading microorganisms, the isolates reported were predominantly aerobic (Table A.2). Of all hydrocarbon degrading isolates (Tables A.1 and A.2) 83 % were aerobes, 13 % anaerobes, and 4 % grew aerobically and anaerobically. Of all oilfield isolates (including non-hydrocarbon degraders), only 3 % grew aerobically and anaerobically and the majority (68 %) were obligate anaerobes. Most hydrocarbon degrading microorganisms were isolated from either contaminated or even pristine sites (89 %, Table A.2). Of all hydrocarbon degraders, 37 % used aliphatic substrates (oilfield isolates 58 %), 39 % aromatic compounds (oilfield isolates 4 %), and 7 % used both (oilfield isolates 17 %). The shift towards aliphatic hydrocarbons among oilfield isolates is also reflected in the observation that many oilfields show severe patterns of biodegradation [43, 44]. We define oilfield isolates as microbial pure strains obtained directly from drill cores or drilling equipment as well as from oil–water separators and other equipment directly installed at extraction or injection wells. Surprisingly, aerobes were isolated from oilfields as well (29 %, Table A.1). Anaerobes grew either on aliphatic or aromatic compounds and many aerobes could use both. OIP degrading microorganisms work syntrophically to gain energy and build up biomass. Two types of biomass producing microbial communities inhabit oil reservoirs. One type decomposes oil and allows the parallel development of heterotrophic nonhydrocarbon degraders. Both hydrocarbon degraders and nonhydrocarbon degraders convert their substrates into CO_2 and methane. Another community type comprises autotrophs that synthesize biomass by CO_2 fixation using inorganic energy sources such as H_2, H_2S, and minerals. Again, heterotrophs coexist within the autotrophic community. Typical indigenous species belong to the heterotrophic, hydrogenotrophic, and CO_2 utilizing genera *Thermotoga*, *Petrotoga*, *Thermoanaerobacter*, and *Thermococcus*. Others are methanogens such as *Methanobacterium*, *Methanococcus*, and *Methanoculleus* (Table A.1; Refs. [19, 28]). *Bacillus* has been reported to be indigenous as well [45]. It often remains unclear whether sulfate reducers were indigenous before drilling, albeit detected in some reservoirs [19, 28]. Results obtained from sequencing 16S rRNA marker genes suggest that in producing oilfields, many oil reservoir species are actively involved in sulfur cycling [19, 46–48]. This is in good agreement with isolates obtained from producing oilfields (Table A.1).

2.3 Microbiological Sampling Techniques for Oil Reservoirs

To study the microbiology of oil reservoirs, it is crucial to consider an appropriate sampling strategy. In the majority of the cases, sampling is restricted to production water directly obtained from wellheads or pipelines with the advantage of reducing labor and costs [19, 49]. With this method, however, microorganisms that inhabit production equipment are included in the sample. This is undesired because microbial processes in subsurface reservoirs and not in the production equipment are relevant for MEOR. In oil producing reservoirs, microorganisms were introduced by drilling and flooding. Therefore, the actual microbial community of producing reservoirs is a mix of microorganisms introduced during production operations and indigenous microorganisms present before any human activity. Aseptic sampling methods should be employed but are hardly applicable in producing reservoirs due to economic constraints. Most oilfield isolates were obtained from production equipment such as production wells or oil–water separators (Table A.1).

One way to obtain microbial communities indigenous exclusively to the reservoir is to sample from the production liner as reported by Kotlar et al. [50]. The authors used Xpand pressure flasks that could withstand pressures up to 100 bar for their sample recovery. Another study compared communities of production equipment and indigenous communities in cultivation experiments [51]. The results revealed a shift of microbial activity from sulfate reduction to methanogenesis after a sterilization treatment of the well, indicating that the indigenous community was methanogenic and sulfate reducers were introduced during operation. Also the number of cells at the well bottom decreased from 2.5×10^5 cells ml^{-1} to 10^4 cells ml^{-1}. The well sterilization method has the advantage that subsequent culturing experiments can reveal the actual in situ physiology. However, the use of vast amounts of highly reactive bleach is questionable if employed on large scale.

Although more expensive, direct subsurface sampling of drill cores is the preferred alternative (i) to minimize the contamination risk and (ii) to collect barotolerant microorganisms that live under actual reservoir conditions. Because an immediate pressure decrease can result in cell rupture and therefore in cell death [52], a slow and controlled sample depressurization to atmospheric pressure was recommended [50]. Indigenous microorganisms are then allowed to slowly adapt to the lower hydrostatic pressures. Furthermore, core samples of a virgin oilfield can be taken [45, 46, 53].

Spark et al. [46] differentially assessed the microbial diversity after DNA extraction. They compared drilling mud communities (only introduced microorganisms) with those found in drill cores. Sampling inner core material is an inexpensive standard procedure to obtain pristine core material [54]. Although obtaining drill cores from mature oilfields would give the most reliable results needed for MEOR, it may be uneconomical especially for oilfields that operate at the economic limit. An honest cost–return assessment should be made. Ultimately,

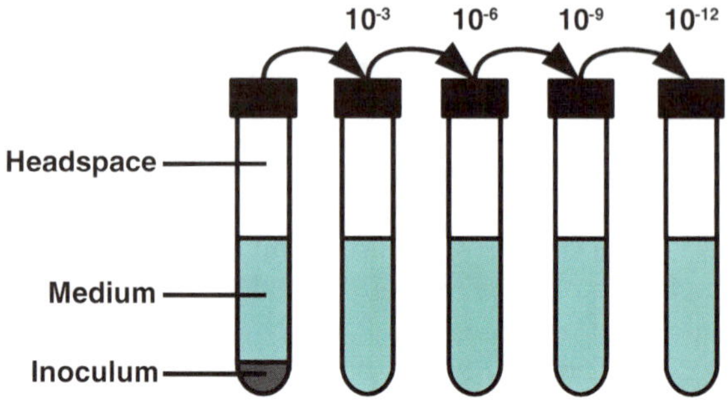

Fig. 3 Dilution-to-extinction method for the isolation and enrichment of living microorganisms in samples from oilfields. The sample is diluted to extinction so that in the last dilution no growth occurs. Knowledge of the original microbial load (cell number) determines the greatest dilution

the community differentiation method of Spark et al. [46] could also be used to distinguish microbial communities within producing oil reservoirs without additional drilling. For example, biosurfactant producing *Bacillus* cells were injected into the Bebee oilfield (OK, USA) and could be recovered at production wells [55]. This experiment demonstrated that, using technically simple methods, intrinsic and introduced microbial populations can be differentiated under production conditions if the introduced population is known.

Whatever the sampling strategy is, it is important to maintain aseptic conditions by using autoclaved or dry sterilized sampling equipment (bottles, tubing, filters, spatula, syringes, etc.) and to keep samples under anoxic conditions. Strictly anaerobic microorganisms can be killed by oxygen due to their lack of protective enzymes [56]. Several anaerobes, like some *Desulfovibrio* species, have adopted such protection mechanisms or even make use of oxygen at low concentrations [57, 58]. However, due to the anoxic nature of oilfields, it is necessary to minimize oxygen intrusion by a nitrogen or argon atmosphere. This atmosphere can be provided in inflatable gas bags (e.g., for opening drill cores) or by a gas flow that covers the sampling area (e.g., for liquid samples). Liquid samples are stored and transported in rubber-sealed and nitrogen-flushed bottles. An anaerobic glovebox is an appropriate alternative to store and proceed on core samples also during field work.

2.4 Characterization of Microbial Populations in Oilfields

2.4.1 Enrichment and Isolation of Microorganisms from Oilfields or Coal Deposits

Basso et al. [51] were the first who thoroughly investigated differences between communities that settled on production equipment and the well bottom. In subsequent culturing experiments, the number of sulfate-reducing microorganisms and heterotrophs decreased whereas hydrogenotrophic methanogens increased, reflecting the nutrient limited conditions in the reservoir fluids. A simple test to associate in situ major microbial groups with physiology is the most probable number (MPN) method. This method determines the abundance of microorganisms capable of growing under certain physiological conditions (e.g., nitrate/

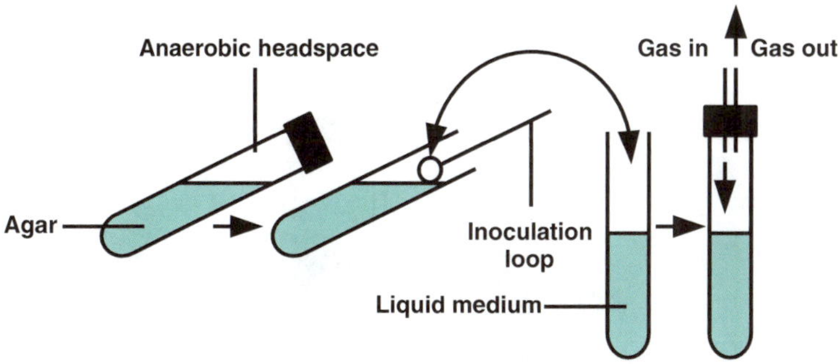

Fig. 4 Isolation technique using tilted 1 % agar tubes. This technique works in both directions and can be used as a complementary method to the dilution-to-extinction method (Fig. 3). Headspace and medium must be anaerobic which is achieved by flushing with oxygen-free gas and by adding an oxygen scavenger to the medium/agar

sulfate reducers). From the abundance of microbial groups linked to physiology one can assess the importance of the investigated pathway. Using MPN in production fluids, aerobic microorganisms can be distinguished as well. Although isolated from oilfields (Table A.1), aerobic microorganisms may be introduced [28]. This is because of the reduced redox state of petroleum reservoirs. The MPN technique is in principle a high-throughput dilution-to-extinction method that can be used
for a broad spectrum of growth conditions (Fig. 3). To better assess microbial physiology, however, rate measurements—by determining reaction products or depletion of substrates—are necessary.

For example, environmental samples are used to inoculate 96 or 384 microwell plates [59, 60]. Anaerobic conditions are maintained by placing plates in an anaerobic glovebox or anaerobic bags [61]. The incubation conditions (e.g., nutrients, salts, temperature, pH, etc.) mimic either the conditions of the injection fluid (e.g., aerobic) or the reservoir (e.g., anaerobic). Other high-throughput instruments for metabolic screening have been commercialized to decrease workload (e.g., Biolog, Hayward, CA, USA). As an alternative to the dilution-to-extinction method, tilted 1 % agar tubes can be prepared to streak out enrichment fluids using an inoculation loop (Fig. 4).

Transfers of isolates can be carried out using a laminar flow bench to maintain sterility under aerobic conditions. If anaerobic microorganisms are to be isolated, the work must be conducted rapidly in order to minimize oxygen penetration into the medium. The headspace should be flushed immediately after closing the tube and an oxygen scavenger may improve the results. Another simple test for the potential microbial life in oil reservoirs is to grow microorganisms on injected or produced fluids [62]. This is achieved by cultivation of fluids in sealed serum bottles or so-called Hungate tubes with an appropriate gas phase (oxic or anoxic mixture of gases) adjusted to reservoir conditions (Fig. 5; [63]).

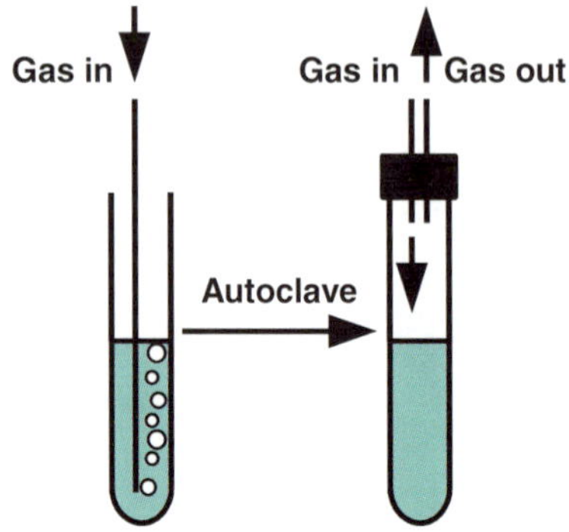

Fig. 5 Preparation of anaerobic enrichment cultures using Hungate tubes [63]. An oxygen-free gas is bubbled through the medium and then autoclaved. After autoclaving, excess oxygen that was still dissolved in the medium is removed by flushing the headspace with a gas that may contain the substrate (volatile hydrocarbons, CO_2). Then the medium is completed by addition of carbonate, phosphate, trace elements, vitamins, and other compounds that are not heat-stable, easily precipitate, or evaporate [67]. Ultimately, oxygen scavengers (e.g., sulfides) may be added to remove trace oxygen

Serum bottles are closed using butyl rubber stoppers and aluminum caps. Complete sets of bottles (50 ml to 1 l) and Hungate tubes (15–20 ml) along with stoppers and aluminum seals can be purchased from various suppliers (e.g., Bellco Glass, Chemglass, Wheaton, Geo-Microbial Technologies, etc.). For cultivation of larger volumes (250–2,000 ml) bottles of 45 mm orifice diameter may be used that are sealed with black butyl rubber stoppers (Wheaton, Millville, NJ, USA) and tightened with heat-stable (up to 180 °C) screw caps with holes (Corning, Tewksbury, MA, USA). All rubber stoppers should be cleaned using soap, acid, and by autoclaving them in distilled water. This procedure should be repeated with a new portion of water until the water remains clean after autoclaving. This procedure is required for new rubber stoppers. Stoppers may be reused and can be cleaned in dishwashers for laboratory. It is important to use thick rubber stoppers of high quality that are ideally PTFE coated. Uncoated butyl rubber stoppers normally contain compounds that are used for growth by aerobic microorganisms [64]. Moreover, thin rubber stoppers leak out hydrogen gas that is used for methanogenesis. Serum bottles should not be filled up completely with liquids, leaving the headspace for gas phase (50:50 $\%_{v/v}$ or 60:40 $\%_{v/v}$). This gives the possibility for addition of various liquid or gaseous compounds and also for taking gas or liquid samples. Horizontal incubation at reservoir temperature with little shaking (<50 rounds per minute) improves gas exchange. To avoid contact of petroleum compounds with the stopper, it may be necessary to incubate without shaking by positioning bottles in a way that prevents contact of rubber stoppers with the medium [65].

Cultures must be incubated in the darkness to inhibit growth of phototrophs which can produce oxygen or other toxic compounds. For cultivation under anoxic conditions, headspaces must be flushed with sterile (<0.2 μm PTFE or polysulfonate syringe filters) pure nitrogen gas or mixtures of nitrogen and carbon dioxide for about 5–10 min (Fig. 5). Passing the gas through copper chips removes trace oxygen [66]. Copper chips can be regenerated in a hot stream of H_2. Gas manifolds (Northern Brewer, Roseville, MN, USA; Glasgerätebau Ochs, Bovenden, Germany) facilitate the preparation of multiple reactors in parallel. Depending on the amount of bicarbonate dissolved in the analyzed fluids, the proportion of carbon dioxide in the mixture (to buffer the pH) can vary from 10 to 20 $\%_{v/v}$. For detection and enrichment of aerobic microorganisms the gas phase may be 100 % or less sterile air. Using less than 100 % air, oxygen-limiting conditions can be mimicked. Addition of various electron donors (oil, individual hydrocarbons) and acceptors (oxygen as air, sulfate, nitrate, and others) might be useful for estimating possible scenarios for specific physiological groups of oilfield microorganisms for MEOR treatment (stimulation, inhibition). Products of microbial metabolism (gases, dissolved organic and inorganic compounds) are monitored by appropriate analytical methods (gas chromatography, HPLC, colorimetry, etc.).

All additions and sampling are carried out without opening bottles using sterile syringes with needles penetrating rubber stoppers. Sterile disposable syringes with plastic barrels are sufficient for such experiments. It is advisable to avoid syringes with a rubber piston on the plunger whenever possible working with oily fluids. Liquid hydrocarbons are easily stacked on or swell the rubber resulting in inaccuracy and difficulty of the syringe handling. However, syringes of 5 and more milliliter volume with pure plastic plungers are normally not tight enough and should be avoided with high internal pressure, for example, methanogenesis. In such cases it is possible to use syringes with rubber pistons carefully avoiding contact between the rubber surface of the plunger and petroleum compounds. Gas-tight glass syringes with Teflon plunger and Luer lock tips may be used for the addition of hydrocarbons. Suppliers of these syringes are various companies dealing with products for laboratories (e.g., SGE Analytical Science, Victoria, Australia). For larger volumes of removed liquids or gases, these should be replaced by adding an equal volume of sterile gas to maintain internal pressure. Syringes should be flushed with sterile anoxic gas before usage because their "dead volume" contains air. Addition of this amount of air may change redox conditions of the experiments or can be toxic for anaerobic microorganisms such as methanogens [68]. If available, the easiest way is to allow gas exchange of sterile wrapped equipment in an anaerobic glovebox (e.g., Coy Glovebox, Grass Lake, MI, USA; MBraun, Garching, Germany). However, experiments usually do not require expensive equipment such as an anaerobic glovebox. The gassing techniques described above can be established quickly in any laboratory with the help of high-pressure bottles of purified inert gas, double-stage pressure regulators (to reduce gas pressure down to 1 bar or less), and some metal and plastic tubing (oxygen can penetrate silicon tubing) for higher and lower pressures of gases, respectively. To sterilize gas, single-use sterile PTFE-filters (<0.2 μm) or a sterile

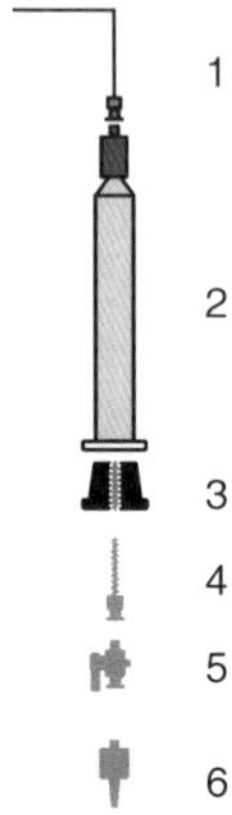

Fig. 6 Assembly of a gassing syringe to be hung in a hot solution: *1* curved needle; *2* 5-ml syringe with Luer lock tip and with a glass barrel filled with cotton; *3* black butyl rubber stopper for closing serum bottles; *4* metal Luer lock female connector; *5* two-way metal stopcock; and *6* Luer lock male tubing connector made from polypropylene is connected to a gas line by plastic tubing

Table 2 Recipes for isolation and growth media modified after Ref. [67]

Fill up to 950 ml water	Freshwater	Brackish	Saltwater	Seawater
KH_2PO_4	0.20	0.20	0.20	0.20
NH_4Cl	0.25	0.25	0.25	0.25
KCl	0.50	0.50	0.50	0.72
KBr	0.09	0.09	0.09	0.09
$CaCl_2 \times 2H_2O$	0.10	0.10	0.15	1.40
$MgCl_2 \times 6H_2O$	0.4	0.4	3.0	5.7
$MgSO_4 \times 7H_2O$	–	0.5	3.4	6.8
NaCl	1	10	20	26
Supplements for all media				
pH	6.8–7.2			
After autoclaving				
Headspace N_2/CO_2	80/20 or 90/10 %			
$NaHCO_3$ (8.4 %)	30 ml			
Trace elements	10 ml			
Vitamins	10 ml			
Optional $NaNO_3$ (1 M)	5 ml			
Na_2S (1 M)[a]	0.5 ml			
Na_2CO_3 (1 M)	1 ml			

Concentrations are g l^{-1} if not indicated otherwise. Note that for the freshwater medium the trace elements solution already contains sufficient sulfate (Table 3). Bicarbonate may be adjusted between 5 and 45 mM final concentration. Here, 30 mM are used. Note that for enrichment of non-sulfate reducing microorganisms sulfate is replaced by an equimolar amount of the chloride salt. Nitrate is added only for nitrate reducers. For mimicking in situ conditions, not more than 1 ml vitamins and trace elements should be added. Mg, Ca, Na, and so on are essential for microbial metabolism and should be added with the trace elements when not supplied with the base medium

[a] An oxygen scavenger other than Na_2S may be used (see text)

Table 3 Possible recipes for trace elements (left) and vitamins (right) added to isolation or enrichment media

De-ionized water	1 l	De-ionized water	1 l
NTA	1.5 g	Pyridoxin × 2HCl	50 mg
or $Na_2EDTA \times 2H_2O$	2.6 g	Thiamin × 2HCl	10 mg
or HCl (25 %)[a]	4 ml	B_{12} (Cyanocobalamine)	10 mg
$MnSO_4 \times 2H_2O$	0.5 g	*p*-Aminobenzoic acid	10 mg
$FeSO_4 \times 7H_2O$	1 g	Riboflavin	5 mg
$NiCl_2 \times 6H_2O$	0.2 g	Nicotinic acid	5 mg
$CoCl_2$	0.1 g	Ca–D(+)-pantothenate	5 mg
$CaCl_2 \times 2H_2O$	0.1 g	Lipoic (thioctic) acid	5 mg
$ZnSO_4$	0.1 g	D(+)-biotin	2 mg
$CuSO_4 \times 5H_2O$	0.01 g	Folic acid	2 mg
$AlK(SO_4)_2$	0.01 g		
H_3BO_3	0.01 g		
$Na_2MoO_4 \times 2H_2O$	0.01 g		
$Na_2SeO_3 \times 5\ H_2O$	0.01 g		
$Na_2WO_4 \times 2\ H_2O$	0.01 g		

To prepare the trace elements solution, NTA or Na-EDTA is dissolved in water by adding KOH until the pH is 6.5. Then the minerals are added and the pH is adjusted to 7 using KOH and the volume is completed to 1 l. Both solutions are bubbled with N_2 and filter sterilized into sterile N_2-flushed serum bottles using <0.2 μm PTFE filters. Note that when modifying the composition, sulfur is an essential element. Ni concentration can be decreased to 0.02 g if methanogens are not of interest

[a] For hydrochloric acid, pH-adjustment is not required

glass barrel of 5 or 10 ml syringe with Luer lock tip, filled with cotton, can be connected to tubing and used for flushing bottles and syringes (Fig. 6). It should be equipped with a bent hypodermic needle, 17–21 gauge [66, 67, 69].

Salinity and pH must be adjusted to reservoir conditions. Often, seawater is injected into depleted oilfields to extract remaining OIP. Because seawater may not be available in the laboratory, artificial seawater recipes have been developed that provide controlled standard conditions (Table 2, [67]). To prepare such media, a stepwise procedure must be followed. First, non-precipitating and heat-stable salts are dissolved in de-ionized water. This usually means that phosphates, carbonates, vitamins, and trace elements (Table 3) must not be added before autoclaving.

Anoxic stock solutions of various substrates can be prepared in serum bottles using a modified vacuum-vortex technique described by Wolfe and Metcalf [70]. Efficient vacuuming can be improved by stirring and solutions become anoxic after three vacuum-stirring steps. For trace elements solutions, non-chelated solutions may be used for sulfate reducers and chelated for nitrate reducers [65]. The chelator nitrilotriacetic acid (NTA) is biodegradable and ethylenediaminetetraacetate (EDTA) is not. Vitamins and trace elements are sterilized by filtration (0.2 μm PTFE syringe filter) and stored refrigerated (4 °C) in the darkness. An oxygen indicator such as resazurine (1 mg l^{-1} final concentration) may be added. The

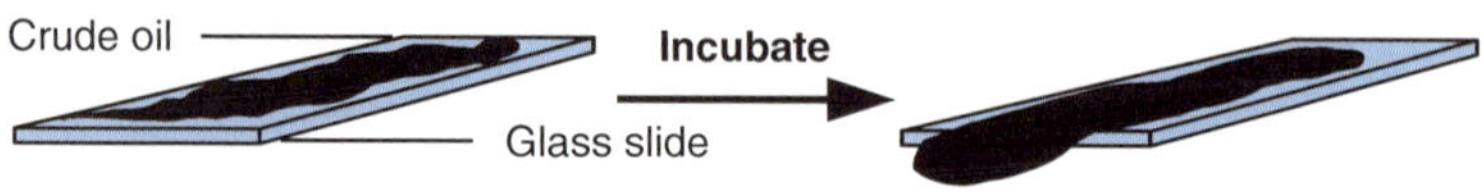

Fig. 7 The glass slide test. Crude oil that has been streaked on the surface of a microscope glass slide will be detached if SAC are produced by microorganisms used as inoculum

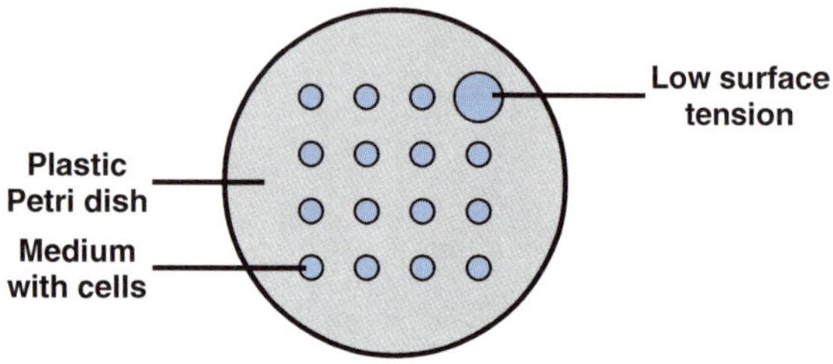

Fig. 8 The plastic Petri dish test. As a result of decreased surface tension, microbial cultures producing SAC have greater wetting areas (drop diameter) on a plastic Petri dish

obtained solution is gassed with argon, N_2, or a mixture of N_2 and CO_2 (usually 80:20 to buffer the pH, Fig. 5).

After sealing and autoclaving, the remaining oxygen has been expelled from the medium into the headspace and is removed by flushing the headspace again with a desired sterile gas mix penetrating the stopper (Fig. 5) or by opening the hot vial in sterile laminar flow and hanging in a sterile gas flow using the outlet as depicted in Fig. 6. Then the medium must be completed by adding filter-sterilized trace elements, vitamins, phosphates, and carbonates to the medium (Table 2). Cellulose acetate filters must be avoided as released cellulose and acetate may act as surfactant or substrate. The pH of the final medium may be adjusted to neutral or slightly alkaline conditions using saturated $NaCO_3$ solutions or to acidity using CO_2 gas. Trace oxygen can be removed using oxygen scavengers such as Na_2S (0.1–1 mM final), $Na_2S_2O_4$ (Na-dithionite; 1 mM final), FeS (prepared by precipitating $FeCl_2$ and Na_2S solutions; 0.5 mM final), Ti(III)-citrate (prepared by mixing Na_3-citrate with $TiCl_3$ and $NaCO_3$-neutralization [71]; 0.5 mM final), dithiothreitol (a carbon source, 1 mM final, [72]) or cysteine-HCl (a carbon source, 10 mM final). It must be considered that some oxygen scavengers add nutrients or surfactants to the medium. Using OIP as a carbon substrate for microorganisms makes the additional supply with another carbon and energy source, such as molasses, unnecessary [62].

Once several enrichment or pure cultures have been obtained, screening for SAC produced by the enrichments or isolates can begin. A simple test is the slide detachment test where crude oil is streaked onto a glass slide used for microscopy, inoculated with a microbial culture and then incubated for several weeks under reservoir conditions (Fig. 7). Cultures that produced SAC detach crude oil from the glass slide and are selected for further investigation. Even simpler but less specific for certain types of crude oil is the plastic Petri dish test. A drop of an enrichment or pure culture is placed on a hydrophobic plastic Petri dish (Fig. 8). SAC decrease the surface tension of the culture medium which becomes visible as

Table 4 Anaerobic microbial processes in oil reservoirs and their respective functional genes

Microbial process	Catalyzed reaction	Functional target gene	References
Nitrate reduction	Nitrate reduction	*narG*, nitrate reductase	[77, 78]
	Nitrite reduction	*nirK/nirZ*, nitrite reductases	[78, 79]
	Nitrous oxide reduction	*nosZ*, nitrous oxide reductase	[78, 80]
Sulfate reduction	Adenosine-5′-phosphosulfate (APS) reduction	*apsA*, APS reductase	Reviewed in [81]
	(bi)sulfite reduction	*dsrAB*, dissimilatory (bi)sulfite reductase	Reviewed in [81, 82]
Methanogenesis	Conversion of methyl group to methane	*mcrA*, methyl coenzyme M reductase	[83–85]
Hydrocarbon degradation	Fumarate addition	*bssA*, benzylsuccinate synthase; *assA*, alkylsuccinate synthase; *nmsA*, naphthylmethylsuccinate synthase	[86]
	Aromatic ring-cleaving hydrolysis	*bamA*[a], 6–OCH–CoA hydrolase	[87, 88]
	Aldehyde:ferredoxin oxidoreduction	*bamB*[a], aldehyde:ferredoxin oxidoreductase	[87]

Note that the presence of one subunit of an enzyme indicates the presence of the entire process
[a] *bam* benzoic acid metabolism

a greater wetting area (drop diameter) on the Petri dish. Using oil-coated glass Petri dishes makes this test even more specific [73].

2.4.2 Identification of Key Microorganisms Using Molecular Biology Techniques

Anaerobic culturing techniques often require special training and equipment and results are sometimes obtained after months or even years. More rapidly, the indigenous community may be characterized by molecular biological methods for identifying major microbial groups. This is important because environmental microbial communities are largely dominated by microorganisms with yet uncultured relatives [74]. The diversity can be addressed by a multitude of well-established methods and reliable kits are commercially available. Most of these methods are based on specific DNA probes that bind to their complementary, evolutionary more conserved regions in genes of interest. Depending on the scientific objectives, one could either choose group-specific genes for certain phylogenetic classification (e.g., 16S rRNA genes to distinguish *Geobacteraceae, Desulfobacteraceae*, *Planctomycetales*, *Methanosarcinales*, etc.) or so-called functional genes. These are genes that encode for enzymes directly catalyzing biogeochemical reactions (Table 4). Functional genes are investigated when microbial populations corresponding to a certain metabolic pathway are of interest (e.g., nitrate reduction, methanogenesis, sulfate reduction, or hydrocarbon degradation). Other genes of interest encode for surfactant producing enzymes, such as the *srfA3/licA* and *sfp* genes [75, 76]. Available methods can be categorized into (i) the enumeration of microorganisms (e.g., by cell staining and microscopy or gene quantification), (ii) the diversity of microbial communities as well as phylogenetic affiliation of sequences (requires DNA extraction and/or microscopy), and (iii) the detection of a metabolically active subpopulation (requires RNA extraction).

A variety of different nucleic acid stains is available on the market (SYBR Green®; DAPI, 2-[4-amidinophenyl]-6-indolecarboxamidine; acridine orange, 3,6-bis[dimethylamino]acridine), that intercalate into the DNA, fluoresce and therefore allow the visualization of cells under a fluorescence microscope [89, 90]. Fluorescence in situ hybridization (FISH) is used to detect group-specific members. In this case, a nucleic acid probe (a so-called oligonucleotide, a short nucleic acid polymer of specific sequence, referred to as "oligo") is labeled with a fluorescing dye. The probe hybridizes only with the complementary target rRNA sequence of a cell mix and the fluorescent tag is than detected by fluorescence microscopy [91]. For the visualization of low copy targets a modified FISH method, catalyzed reporter deposition FISH (CARD-FISH), is suggested, where an enzyme-mediated reaction causes a signal amplification [92].

Polymerase chain reactions (PCR) allow for the amplification of specific DNA regions of extracted total DNA. In principle, DNA probes ("oligos," primers) bind to complementary sequences of the investigated gene (template) that can then be copied to detectable amounts by a DNA-synthesizing enzyme (DNA polymerase). The primer binding resembles the key and lock principle. Quantitative real-time PCR (qPCR) can furthermore reveal DNA copy numbers by the use of standards (DNA extracts of resembling genes) and fluorochrome-tagged probes (reviewed in [93, 94]). For example, qPCR was successfully applied to quantify sulfate-reducing microorganisms in the Dan and Halfdan oilfields' production fluids (North Sea) to monitor the potential for microbially influenced corrosion (MIC; [95]). The abundance of certain microbial groups (e.g., methanogens or sulfate reducers) is a good indicator for the respective microbial processes. qPCR has been applied to identify various microbial groups in soil and marine sediments (e.g., Refs. [54, 96, 97]) but reliable data for different oil reservoirs data are insufficient [98–100]. Although qPCR is reliable, cheap, and well established, another recently developed approach not using polymerase enzymes gives the same quantitative information. This NanoString® method also works with heavily degraded mRNA or formaldehyde fixed environmental samples [101].

Microbial diversity analyses further provide information about bacterial and archaeal species present in the sample. Today, usually two DNA sequencing methods are used to determine the nucleotide sequence within a DNA molecule. Sanger sequencing is inexpensive (approximately US $250 per sample returning 100 sequences) but labor intensive (2–3 weeks for laboratory and computer work)

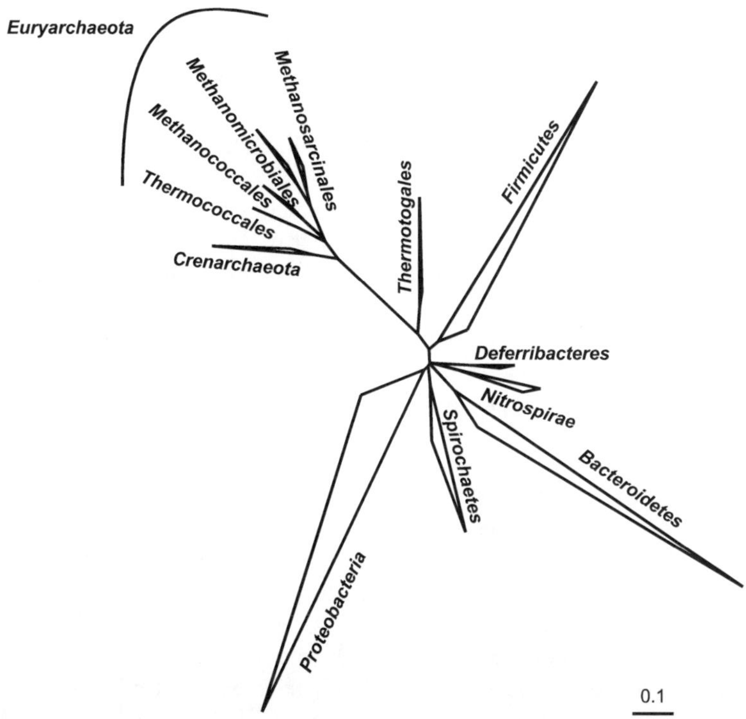

Fig. 9 Phylogenetic tree obtained from the Silva SSU Ref database version 95 of which all uncultured strains were removed. Additionally, all groups in which no oilfield clones were detected were removed as well. Oilfield clones were taken from Refs. [19] and [7]. The bar indicates 0.1 change per 16S rRNA nucleotide

and requires a cloning procedure, during which genetically modified organisms are produced. In contrast, pyrosequencing is less expensive (approximately US $120 per sample), less laborious, and gives in-depth results (up to 5,000 sequences for one sample). In the past, sequence lengths often did not reach 400 base pairs which are necessary for reliable phylogeny on genus level [102] as opposed to Sanger sequencing which routinely reaches >900 base pairs. Samples for pyrosequencing can be combined, reducing the price and the number of sequences obtained. Pyrosequencing is also quantitative as it is not biased by intermediate PCR reactions prior to the actual sequencing PCR. Genetically modified organisms are not produced. Using sequencing techniques, it is possible to assign many sequences to a phylogenetic group by incorporation into a phylogenetic tree (Fig. 9).

A sequencing approach that also addresses functional diversity is the construction of a metagenome (DNA) or a metatranscriptome (RNA). Metagenomic analyses became popular in recent years due to decreasing sequencing cost. Unfortunately, this drop did not trigger a comparable metagenomic flood for oilfield microbiology, possibly because gene annotation still requires hand-picking [42]. For a metagenome, the same problem as for any other DNA targeting method occurs. DNA samples include fossil and therefore inactive DNA as well and can make false predictions. Or, the random primers used do not include present genes as was the case for the predicted absence of nitrate reduction from a metagenome [103], which was then observed using cultivation techniques [104]. This example shows that complementary methods should be employed to achieve a certain level of confidence. Sequencing costs are permanently dropping and quick community identification can precede further culturing experiments. New sequencing techniques such as Illumina sequencing may further reduce sequencing costs [105].

Widely used in biomedical research are microarray techniques where known nucleotide probes are attached to a high-density microarray. Target-probe hybridization can be detected by fluorescence signals. A number of applications exist for phylogenetic 16S rRNA (gene) and functional gene levels. Such arrays were indeed used for environmental samples in recent studies (e.g., Refs. [106, 107]).

Using denaturing gradient gel electrophoresis (DGGE; [108]) or terminal restriction fragment length polymorphism (T-RFLP; [109]) it is possible to compare "fingerprints" of microbial communities in petroleum affected sites [110]. Gene amplicons are needed and profiling is either based on variable content of the nucleic acid guanine that binds with cytosine (G/C-content) for DGGE or the position of restriction sites (DNA/RNA regions that bind highly specific restriction enzymes) closest to the labeled end of the amplicons for T-RFLP. Although inexpensive, these methods can only give information about the diversity. Information about the identity must be interpreted with care as nucleotide sequences are not produced or may not be representative.

Characterizing microbial communities using 16S rRNA on the DNA or RNA level is insufficient for identifying potential biogeochemical reactions (with the exception of methanogens) because substrate utilization is independent of rRNA. For an oilfield operator, however, it is of interest to stimulate or suppress metabolically active subpopulations (e.g., methanogens, sulfate reducers, hydrocarbon degraders, biosurfactant producers, etc.). Marker genes, such as *srfA3/licA* and *sfp*, are known to encode for biosurfactant-functioning glycol- and lipopeptides [75, 76]. Because RNA is rapidly degraded, it is characteristic for active cells. RNA is used as a template in reverse transcription PCR (RT-PCR) for the synthesis of complementary DNA (cDNA, which is more stable than RNA) by reverse transcriptase enzymes and DNA polymerases that use RNA as a template. In a subsequent PCR, cDNA can be used as a template for all methods described above. Alternatively, stable isotope probing (SIP) links microbial identity to function [111]. Added ^{13}C-labeled substrates are incorporated into cell components such as DNA. After nucleic acid extraction, isotopically heavy ^{13}C-containing fractions are separated by ultracentrifugation using a density gradient. Using SIP, microorganisms were identified in hydrocarbon degrading food webs [112, 113] or hydrocarbon seeps [114]. Furthermore, other techniques including spectroscopy, spectrometry, or radiography are combined with molecular detection tools: (i) FISH-microautoradiography, FISH-MAR, (ii) FISH-Raman microspectroscopy, FISH-RAMAN, and (iii) FISH-secondary ion mass spectrometry, FISH-SIMS (reviewed in [115]). All methods visualize and quantify incorporated isotopes and/

or radioactive signals after feeding isotopically labeled substrates in single-cell resolution. However, these techniques require highly skilled personnel, are labor-intensive and costly (equipment, isotopes, supplies, etc.). All nucleic acid-based methods are restricted to published sequences, therefore only sequences analogous to known species can be detected. New branches of unknown unaffiliated sequences cannot be reliably classified if isolated and described pure strains do not exist. SIP can help to overcome this problem as it relates substrate degradation to 16S rRNA gene sequences, and thus to the identification of the responsible organisms.

Alternatively, lipid biomarkers can be used. Phospholipid-derived fatty acids (PLFA, cell membrane components) are used to estimate microbial biomass and can be assigned to certain microbial groups [116]. Nonetheless, phylogenetic resolution is limited and correlation to underlying microbial processes is often highly speculative compared to nucleic acid-based methods.

In principle, all of the afore-mentioned techniques work well for many environmental samples. However, oil samples confront the investigator with specific challenges and requirements. Deep subsurface microbial habitats that are often hot and depleted in electron acceptors are characterized by relatively little living biomass [23, 54]. Cell counts ranged from 10^4 cells g^{-1} in the oil-saturated zone to 10^7 cells g^{-1} in the oil–water transition zone of a core taken from a Canadian oilfield [117]. Low DNA or RNA extraction yields affect the efficiency of molecular methods. For production fluids, large volumes can be concentrated to enrich microbial cells on <0.2 μm filters prior to DNA extraction. In contrast, cell accumulation by centrifugation is not recommended in particular for high-salinity samples because higher mass densities lead to loss of microbial cells. Further enrichment of genomic DNA may be achieved after DNA extraction by sodium acetate precipitation [118]. However, because additional precipitation may be incomplete, subsequent quantitative methods should be critically evaluated. Qualitative investigations will not be affected.

Only little information for DNA extraction directly from the oil phase or the oil–water interface is available [119]. Extraction of microbial cells from oil can be performed using organic solvents, such as *n*-hexane, methanol, or isooctane [120]. This depends on oil maturation [121]. A second method for DNA extraction uses buffer systems, such as the Winogradsky solution [122, 123]. Alternatively, microorganisms can be extracted from the oil into the water phase followed by filtration and DNA extraction of the residue [124]. Again, an underestimation of extracted DNA may be the result of additional extraction steps. Recovery experiments by using control strains (as internal standard) are recommended for quantitative methods.

For coal, the material is ground up and extraction can be improved using bead mill. Prior to grinding, maceration with 100 mM phosphate buffer might improve results [125]. Alternatively, a combined splitting and drilling method can be used to ensure an uncontaminated sampling surface [126]. Otherwise coal surfaces are cleaned with ethanol or bleach. Subsequent DNA extraction can be carried out using commercially available kits for soil or sediment samples.

Crude oil typically fluoresces in the range of 300–600 nm depending on its carbon structure and additives [127]. Fluorochromes, as used in molecular biology, have an excitation signal maxima ranging from 390 to 740 nm [128]. In consequence, remaining oil components in the extraction eluates may autofluoresce and, therefore, interfere with the detection signal of the excited fluorochrome emission.

3 Selecting an Appropriate MEOR Approach

Among the five different MEOR strategies mentioned in the introduction, the alternation of surface or oil–water interface properties (to release OIP as it is) and methanogenesis (where OIP is converted into natural gas, methane) seem to be the most promising approaches. In the first part we introduce methanogenic hydrocarbon degradation by which OIP or coal components are consumed to form methane. This is the easiest way to recover fossil energy from non-extractable OIP but the product methane is cheap. Methanogenesis may be stopped at some point when oil is swollen enough to recover more valuable components as well. If the degradation of OIP is not desired, more complex modifications of the oil–water interface properties are required which are introduced in the sections covering biosurfactant production, viscosity alteration, and pore space plugging.

3.1 Conversion of Oil or Coal into Natural Gas (Methane)

Microbial conversion of underground carbon (oil, coal) to methane appears to be the easiest way to extract energy from such deposits because energy can be captured from the gas phase. Fermentation and methanogenic processes have been proposed as strategies to enhance energy recovery from stranded energy assets (i.e., reservoirs where over 70 % of the resource had to be left in place due to extraction limitations) by stimulating microbial activity. Methanogenic reactions are exergonic (Table 1), therefore they come at the price of energy loss for biomass accumulation. Several European and North-American countries as well as China are investigating microbiological conversion of residual crude oil and coal into biogenic methane to extend the lifetime of their existing fossil fuel resources. Today's proven oil reserves are estimated at 1,200 billion barrels (Gbbl) based on a mean recovery of 35 % of OIP (roughly a 40-year supply based on current global consumption of 85 Mbbl day^{-1}). Increasing the recovery rate to 50 % would produce incremental 520 Gbbl, extending production from current reserves at current consumption rates by <17 years. Sustaining energy production must, however, go hand in hand with emissions mitigation. This will involve strategies such as carbon capture and storage but can also be facilitated by a switch to lower emission fossil fuels. For every kWh of energy produced, methane generates 0.569 kg of CO_2, whereas oil produces 0.881 kg of CO_2 and coal 0.963 kg of CO_2

[129]. Stimulating the microbial conversion of oil to methane could therefore reduce CO_2 emissions per kWh by one-third. This holds true under the premise that CO_2 generated in the conversion process would not be released into the atmosphere.

Major world oil reserves, such as the Athabasca oil sands, other foreland basins, as well as many offshore reservoirs, contain heavily biodegraded crude oil [130]. The occurrence of biodegraded oil is indicative of indigenous subsurface microbial communities. Therefore, in situ biodegradation in petroleum reservoirs is a globally significant biogeochemical process. Removal of aliphatic and aromatic hydrocarbons during in situ biodegradation enriches heavy oil fractions containing heterocyclic sulfur-, oxygen- and nitrogen-rich compounds [43]. It severely decreases oil quality. Factors controlling in situ biodegradation and the specific microorganisms responsible remain poorly understood. Geological conditions limit aerobic catabolism due to negligible contact between meteoric oxygen-rich waters and deep fossil fuel deposits. The discovery of anaerobic hydrocarbon degradation in various laboratory microcosm experiments (e.g., Refs. [131–136]) and field-scale evidence from biodegraded reservoirs [30, 43] support the hypothesis that anaerobic processes are responsible for subsurface crude oil and coal biodegradation. Inasmuch as electron acceptors for anaerobic metabolism such as nitrate and oxidized metal species (Fe[III], Mn[VI]), are largely absent or sequestered (e.g., as iron silicates), sulfate reduction and methanogenesis are the most relevant processes for in situ biodegradation of crude oil. Sulfate can occur naturally in formation waters of reservoirs rich in soluble evaporite minerals (e.g., halite) or when seawater is injected. Such sedimentary minerals form as a result of dehydration during the formation of hydrocarbons. Methanogenesis is predominant in the absence of sulfate. Because methanogens cannot directly decompose complex organic matter they are dependent on symbionts that produce hydrogen, C_1 compounds, or acetate [137–139]. Substrates degraded by sulfate reducers are more divers and include hydrocarbons. Sulfate reducers can therefore provide intermediates needed by methanogens. It was shown that either inoculation [140] or stimulation [141] might be feasible approaches to produce methane from hydrocarbons. In the inoculation experiment, the authors used a preincubated hydrocarbon degrading methanogenic mixed culture to inoculate sandstone cores of a mature oilfield in Oklahoma [140]. They observed methane formation rates from OIP between 3.4 and 9.0 $\mu l_{CH_4}\ g_{oil}^{-1}\ day^{-1}$. In the stimulation experiment, naturally enriched hydrocarbon degrading communities of contaminated harbor sediment were used [141]. The authors could show that the stimulation of the naturally enriched community led to methane formation rates of 2.5 $ml_{CH_4}\ g_{hexadecane}^{-1}\ day^{-1}$ when ferrihydrite was added compared to 1.1 $ml_{CH_4}\ g_{hexadecane}^{-1}\ day^{-1}$ with sulfate. However, the stimulation study also showed that naphthalene (a polyaromatic hydrocarbon, PAH) was degraded three orders of magnitude slower than *n*-hexadecane or ethylbenzene and that electron acceptors did not accelerate this process. Hence, the relatively small proportion of *n*-alkanes in the inoculation experiment did not affect the overall oil degradation rates and might be the reason for the observed lower rates [140]. It remains unclear how PAH degradation could be accelerated as the microorganisms and degradation processes are poorly understood. Aromatic ring reduction seems to be the overall rate-limiting step [142]. Another way to accelerate methanogenic hydrocarbon degradation is the addition of trace elements and vitamins [143].

Assuming that light crude oils consist of approximately 10–15 % of *n*-alkanes or mono-aromatics (benzene, toluene, ethylbenzene, and xylenes; BTEX), the microbial degradation of this fraction would convert 50 % of these into methane. Hence, approximately an additional 5–10 % of the total oil mass could be recovered. Thus, the induced conversion of oil or coal into methane can increase the production lifetime of these reservoirs. Additionally, the conversion of oil hydrocarbons into methane may have beneficial side effects such as a change of oil viscosity or oil swelling. This is further discussed in Sect. 3.2.

This example of MEOR will depend on a better knowledge of syntrophic microbial consortia in the reservoirs. A better understanding of the factors affecting microbial communities in petroleum systems will have the benefit of potentially resolving some of the uncertainties associated with earlier MEOR attempts. Culturing or molecular biology approaches are appropriate and have been discussed in Sect. 2.4.

3.2 Oil Swelling by Microbial Production of Gases

Microbially produced gases can swell oil by decreasing its density and increasing its volume, and thus facilitate oil displacement [144]. The in situ production of gases to swell OIP may leave a fraction unaltered. It can therefore be an interesting alternative to the complete conversion of OIP to methane while being easier to control than the production of SAC. In recent years, CO_2 was successfully used to swell and recover OIP upon cycled injection with N_2 [145]. However, the main mechanism behind CO_2 injection as EOR method is the displacement of OIP adhered to minerals by CO_2 binding [146]. In addition to CO_2, methane is produced by microorganisms inhabiting oil reservoirs. Methane has a solubility of 0.10 ml ml^{-1} brine at 11 bar and 45 °C [147], 0.24 ml ml^{-1} at 49 bar, 50 °C and 0.84 ml ml^{-1} at 113 bar, 50 °C [148] showing an almost linear increase with pressure. Four milliliter methane dissolve in 1 ml gasoline at 8.4 bar [149]. The higher methane solubility in gasoline compared with brine was responsible for a volume increase of the gasoline by 1 %. It seems unlikely that swelling by dissolved methane at such pressures could facilitate oil displacement. However, under high-pressure reservoir conditions, oil swelling might drive oil out from pore space due to a higher solubility of methane. The formation of methane hydrate in the water phase can be neglected due to the low temperatures needed for hydrate formation. An undersaturated reservoir with a bubbling point of 58 bar (835 psia) and a pressure of 386 bar (5,600 psia) would be favorable for MEOR by gas production [144]. How methanogenesis can be stimulated is described in Sect. 3.1.

To avoid complete destruction of OIP by methanogenic consortia, methanogenesis may have to be inhibited. This can be achieved by using inhibitors such as methyl fluoride (CH_3F, 1.0–1.2 %) for acetoclastic methanogenesis or 2-bromoethanesulfonic acid (BES, 8 μM–5 mM) for general methanogenesis [150]. Oxygen [68] and nitrate [141] also inhibit methanogenesis but accelerate hydrocarbon degradation. Many facultative anaerobic hydrocarbon degrading strains listed in Table A.2 are also nitrate reducers.

N_2 is produced by nitrate reducers. This could increase reservoir pressure, as suggested in Refs. [151–153]. At the same time, reports of increased actual reservoir pressure upon MEOR treatment are scarce. In a review, Belayev et al. [151] mentioned an increase of well pressure (top of the well) in a well stimulation experiment of the Sernye Vody oilfield, Russia, by 1.5 bar. It seems unlikely that this pressure increase could be maintained throughout a reservoir where injection and production wells are separated. Moreover, the small pressure increase is insufficient and cannot account for oil displacement. In a study of an example oil reservoir in the North Sea the increase of reservoir pressure (not oil swelling) was regarded as irrelevant for MEOR due to the formation of a gas phase needed for pressure generation of several orders of magnitude [144]. This gas phase would then potentially block sweeping fluids.

3.3 Detaching Residual Hydrocarbons from the Mineral Matrix (Capillary Number N_{CA})

For petroleum engineering, the ability of microorganisms to remove adhered hydrocarbon from the mineral matrix (i.e., the porous medium that contains petroleum) offers remarkable opportunities. In particular, microorganisms are able to reduce the force that retains hydrocarbons in microcapillaries of the porous medium. This force is best described as the capillary number N_{ca} [154]:

$$N_{\mathrm{ca}} = \frac{\mu_{\mathrm{w}} \times v_{\mathrm{w}}}{\gamma_{\mathrm{ow}}} \tag{1}$$

The viscosity μ_{w} is the viscosity of the water phase and v_w is its volumetric flux. The interfacial tension γ between oil and water is γ_{ow} and the resulting capillary number is dimensionless. In most reservoirs, the capillary number is around 10^{-7} [144]. An increase of N_{ca} by three orders of magnitude is required to displace oil from capillaries of 10–100 μm dimensions [155]. This cannot be achieved by increasing the water viscosity or the volumetric flux. Therefore, a three orders of magnitude decrease of γ_{ow} will displace entrapped oil from pore space. Gray et al. [144] reported some typical γ values for hydrocarbons and water and found that in 1 out of 16 cases γ was decreased below 0.06 mN m^{-1} by microorganisms and their products at neutral pH. Others reported biosurfactants to decrease γ_{ow} by up to five orders of magnitude using the spinning drop measurement method [156].

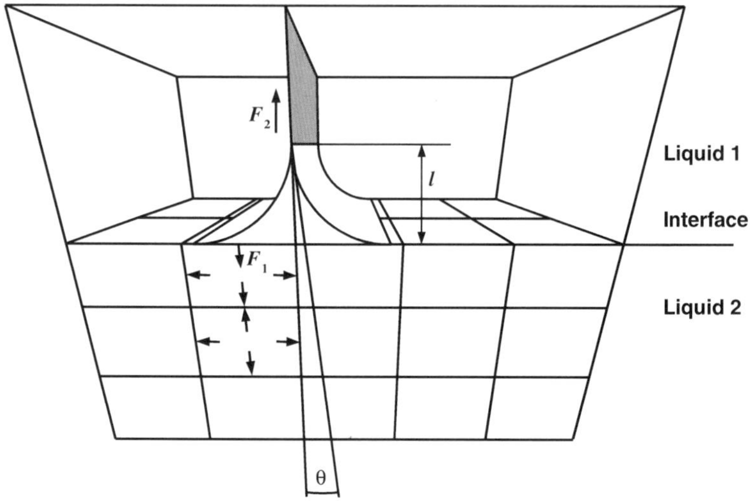

Fig. 10 The Wilhelmy plate method. A liquid body can be imagined as cubes that are being held in place by repulsive and attractive forces. When these forces are balanced (e.g., in a hydrophilic liquid) no phase separation occurs. When these forces are imbalanced, phase separation becomes visible and the difference between these forces can be measured using the Wilhelmy plate. This difference also determines the interfacial tension γ. Note that l is the wetted perimeter of the Liquid 2-wetted area

Because such low γ values were rarely reported, thorough investigation of oilfield microorganisms regarding their interfacial action is inevitable for any MEOR treatment aiming to detach OIP. Detaching OIP is less destructive than conversion to methane but more difficult to control. In the following, we introduce basic techniques that help to address this topic.

3.3.1 Determination of the Interfacial Tension γ

In order to increase N_{ca}, biosurfactants are used to decrease γ_{ow} (Eq. 1). Generally, γ can be measured using different techniques: the Wilhelmy plate method, the pendant drop method, the spinning drop method, and the laser scattering method. To choose between the different tensiometers it is useful to understand their basic functioning (for reviews see Refs. [154, 155]). The Wilhelmy plate method is the classical method and allows for the measurement of $\gamma \ll 1$. In principle, a metallic plate is suspended vertically in a two-phase liquid (Fig. 10).

The force F that is applied to this plate can be measured directly and is used along with the contact angle θ and the perimeter l of the wetted area to calculate the interfacial tension between both immiscible liquids:

$$\gamma = \frac{F}{l \cos(\theta)} \qquad (2)$$

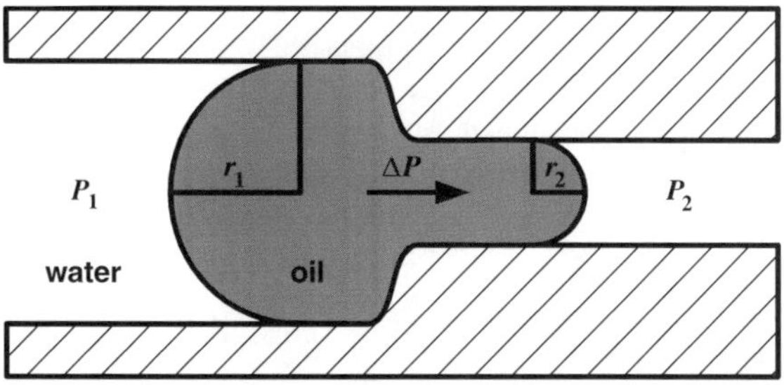

Fig. 11 Model of oil displacement through confined pore space. The interfacial tension is defined in the Laplace–Young equation (Eq. 3) and determines the pressure difference ΔP

The disadvantage of this method is that relatively large amounts of both liquids are needed. The pendant drop method is easy to handle and reliable. The two phases are the drop that is inflated into the surrounding medium (Fig. 12). In addition to the easy handling, this method also allows for incubating an undisturbed surface, for example, to observe the formation of biofilms or to measure the elasticity [157]. The principle of drop tensiometers (pendant drop or spinning drop) is based on measuring of the drop curvature deformation as a result of γ changes ($\Delta\gamma$) using the Laplace–Young equation for capillary pressure [155]:

$$\Delta P = P_1 - P_2 = 2\gamma\left(\frac{1}{r_1} - \frac{1}{r_2}\right) \tag{3}$$

The difference between the two pressures P_1 and P_2 on each side of the drop, ΔP, directly correlates with γ, depending on the difference between the corresponding radii r_1 and r_2 (Fig. 11). That is, the greater the difference between the two radii, the greater the pore constriction, the less ΔP depends on γ. When both radii are equal (the porespace is not confined), only γ determines ΔP.

Ideally, an oil drop in water has a perfectly spherical shape because the pressure of the liquid of the higher density holds the other liquid in shape like a water drop in air without gravity. When the interfacial tension γ is the greatest, the oil drop will have an ideal spherical shape of the lowest surface area and the sinus of the contact angle at the full radius $\theta = 90°$ will be 1:

$$2\pi \times \gamma \sin(\theta) = V(\rho_o - \rho_w) \times g + \pi x^2 P \tag{4}$$

V is the known volume of the drop, ρ is the known density of either oil (ρ_o) or water (ρ_w), g is the known gravity, πx^2 is the area of the drop of the diameter x, and P is the pressure between the drop and the surrounding water. This pressure difference can be assumed to be the same between the water and the oil drop and will therefore be 0. Changing γ will lead to a measurable change of the contact angle θ:

$$\gamma = \frac{V(\rho_o - \rho_w) \times g}{2\pi \times \sin(\theta)} \tag{5}$$

Oscillating the drop volume V using a certain frequency f will lead to a change of the area A, the contact angle θ, and the interfacial tension γ (Fig. 12). The

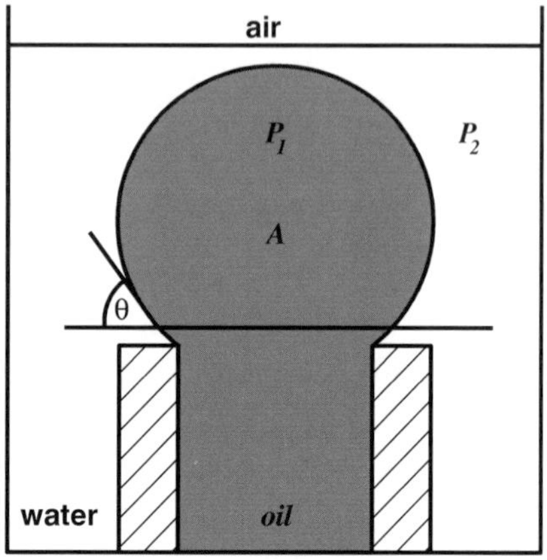

Fig. 12 Principle of the pendant drop method. The oil drop is inflated from the needle tip at bottom into the aqueous phase. Using the contact angle θ and the surface area A, the elasticity E or the interfacial tension γ can be calculated using Eq. (6) or (5), respectively

derivation of the oscillating A and θ ultimately allows the calculation of the elasticity E using Eq. (6). This is important to consider because elasticity can diminish the effect of γ reduction (Sect. 3.3.2).

Briefly, one can easily imagine that changing the oscillation frequency f will lead to a change of E, given there is elasticity. Moreover, f influences the elastic "response" of the drop that is covered by an elastic skin. When f is high, the inflation speed is high along with the difference in pressure and the force that is applied on the surface. This results in little elastic "response" because the elastic forces on the surface of the drop, withstanding the applied forces, are smaller than the applied forces. The result is a shorter delay of the elastic response, measurable as a shift in the phase angle α between the oscillations applied by the inflating instrument (e.g., a pendant drop tensiometer) and the real in- and deflation frequency of the drop.

However, using a pendant drop tensiometer one can measure only relatively high γ values (usually not lower than 1 mN m^{-1}). This is because the oil drop easily detaches when γ becomes too low due to the density difference and the drop migrates to the surface. Unfortunately, γ values above 10^{-2} mN m^{-1} are not sufficient for oil displacement [155]. To solve this problem, a variation of the pendant drop method is the spinning drop method. Here the Laplace–Young equation is used as well (Eq. 3). The costs for a spinning drop tensiometer can vary between US $20,000 and 40,000. It is more expensive than a pendant drop tensiometer but values as low as 10^{-6} mN m^{-1} can be measured. This is achieved by a freely floating oil drop within the aqueous medium. The medium rotates with a drop inside a capillary and the injected drop does not contact the boundaries of the capillary. The spinning drop is extremely deformed so that studying an undisturbed surface becomes impossible. Moreover, γ cannot be measured continuously without rotation because a new drop must be injected each time the measurement has been halted. A novel technology that combines the features of an undisturbed surface with the measurement of low interfacial tensions is laser scattering tensiometry (SINTEF, Trondheim, Norway). Laser scattering instruments are usually employed to measure the shape and size of particles and can be used for oil droplets as well. The lowest γ measured using laser scattering is 10^{-2} mN m^{-1}. However, this technology is still under development and not commercially available yet.

3.3.2 Impact of Oil Droplet Elasticity: Gel Formation

The production of SAC that decrease γ is naturally accompanied by an increase of elasticity E [158]. Consequently, higher pressure must be applied for droplet deformation and oil displacement. The mechanism of droplet elasticity is not well understood and the behavior of stiff oil droplets is not predictable yet [144]. One can picture surfactants as coating oil droplets, determining the droplet's elasticity. When a drop is deformed during its passage through a pore constriction, by applying pressure on one side of the drop in order to displace it, the surface area A of this drop will change along with its γ. The relation between γ and the resulting pressure difference ΔP is described in the Laplace equation (Eq. 3). The radius r_1 of the drop curvature is greater than r_2 on the confined (right) side of the pore constriction (Fig. 11). It is obvious that when γ becomes very low, that the difference in pressure ΔP decreases, thereby facilitating drop displacement. At the same time, γ is directly linked to E. An increase of the drop elasticity (i.e., interfacial relaxation time) could impede drop displacement. This is because the elasticity of a drop, covered by a skin of surfactant molecules, depends on the change of γ (dγ) when the area of the drop changes (dA), for example, during in- and deflation of the drop [159]. This dependence is expressed in the Gibbs formula:

$$E = \frac{\mathrm{d}\gamma}{\mathrm{d}\ln(A)} \tag{6}$$

where E is the elasticity, γ is the interfacial tension, and A is the area of the drop that is deformed. The unit of the elasticity is the same as of γ, N m^{-1}. For example, when ln(A) increases by 0.01 due to the expansion of the drop surface on the confined side of the drop (the right side in Fig. 11) at a given elasticity of 50 mN m^{-1}, γ will increase by 5 mN m^{-1}. Then the increase of γ constrains drop mobility although a surfactant had decreased γ before. In conclusion, using microbial cells to study their effect on γ may have nothing to do with the actual γ when these cells produce biosurfactants or form a surface biofilm [144]. This is because a coated oil drop does not have an interface with water when cells grow on its surface.

Crude oil naturally contains compounds having the same effect on the interfacial tension and the elasticity [158]. These compounds may have a polar head group and a lipophilic tail as well, for example, naphthoic acid. Because these compounds are characterized by a great variety, one can assume that they form patterns or patches on the drop surface. These mutually interact, ultimately resulting in the formation of a fractal gel covering the surface of an oil drop [160]. Gels are impermeable for other compounds and deforming the drop will not change its permeability. This could lead to stable oil–water emulsions. Such emulsions impose economic risks to oilfield operators because it becomes difficult or even impossible to decrease γ or separate oil from water. Gels of intrinsic surfactants increase the relaxation times of oil drops and therefore may impede their displacement from the porespace. Due to the heterogeneous nature of the intrinsic oil surfactants, the elastic response is the same at all oscillation frequencies. In other words, a wide range of frequencies will not break gels. This can indeed be measured as a relatively constant phase angle at different oscillation frequencies [159]. Increasing the hydrophobicity of the aqueous phase or altering the viscosity may affect droplet elasticity. Indeed, detachment of cells was observed when 2-propanol was added to a cell-coated oil hexadecane suspension [161]. Other amphiphilic solvents may have a similar effect on oil–water emulsions. Some typical organic solvents produced by microorganisms are, for example, acetone and butanol, excreted by certain *Clostridia* [162].

3.3.3 Biosurfactants and Other Surface Active Compounds in MEOR

Surfactants are amphiphilic molecules that accumulate at interfaces, decrease γ, and form micelles [163]. They can be classified by properties such as their critical micelle concentration (CMC), their ratio of the hydrophobic to the hydrophilic regions (hydrophile–lipophile balance), their chemical structure, and charge, among others. Due to these properties, surfactants affect the way molecules interact with interfaces. In natural ecosystems most microbial activities occur at interfaces that may be liquid or solid. Hence, it is not surprising that surfactant synthesis (then called biosurfactants) is common among microorganisms (for reviews see Refs. [163, 164]). There is a great variety of chemical structures because biosurfactants can be peptides or esterified lipids. Biosurfactants coat surfaces by forming a molecular monolayer skin and decrease γ. Naturally, these biosurfactants serve various functions, such as cell motility, attachment to, or detachment from surfaces, biofilm formation, pathogenesis, antimicrobial activity, modification of cell surface properties, ingestion of hydrophobic substrates, and so on.

A classic example for a biosynthetic SAC that is not a biosurfactant is acetone. It is soluble in water and in benzene. Its short lipophilic tail does not allow the formation of an elastic skin, characteristic for biosurfactants. This is also the case for other short-chain alcohols and ketones typically produced by microorganisms [165]. Fatty acids may be further esterified with hydrophilic head groups which often increase their solubility in water. Typical hydrophilic head groups are carbohydrates, peptides, phosphates, and sulfates or a combination of them. Some powerful biosurfactants including their γ are listed in Table A.3. The most important biosurfactants have been reviewed by Banat et al. [166] and more recently by Mulligan [167] and Rosenberg [164]. In brief, well-described biosurfactants are: surfactin (a lipopeptide) of *Bacillus* species as well as viscosin (lipopeptide) and rhamnolipids (glycolipides) of *Pseudomonads*. *Bacillus* sp. biosurfactants could decrease γ to <0.1 mN m^{-1}, needed to mobilize oil in suspension [168], in micromodels [169], and in an oilfield [55]. However, in another field study of the Changqing oilfield in China γ was reduced to values greater than 10 mN m^{-1} in four out of six inoculated wells. This was insufficient for substantial additional oil displacement, showing that an extrapolation from the lab to the field is not always possible [170].

Bacillus sp. also produces high-molecular-weight compounds that may act as biosurfactants [171]. Generally, such compounds are polymers of various origin including carbohydrates, nucleic acids, proteins, and lipids. These, when excreted by cells, form extracellular polymeric substances (EPS, [163, 164]). EPS are the building blocks of biofilms that hold microbial cells attached to their organic or inorganic substrates. A well-studied polysaccharide is the emulsifier RAG-1 which is produced by several species of *Acinetobacter* (Table A.3; [172]). RAG-1 was able to stabilize an emulsion of aromatic and aliphatic hydrocarbons.

Useful tools to study the action of polymers in the lab are micromodels. Micromodels are two-dimensional artificial porous media that help visualizing loading and unloading the porespace with any liquid [33]. The fabrication of a micromodel costs usually less than US $1,000 and additional equipment for high-pressure injection is commercially available. Micromodels are therefore a good alternative to core flooding experiments especially because the process can be visualized.

It was shown in numerous reports that (bio-)surfactants facilitate hydrocarbon degradation [169, 173–175], even when they were synthesized by a nonhydrocarbon degrading species [176]. Biodegradation may be desired (to produce biosurfactants) or not (to maintain oil quality). However, that accelerated biodegradation of OIP can be facilitated by biosuractants is not a general rule. For example, the biodegradation of crude oil by a mixed consortium of *Enterobacter* and *Pseudomonas* was not significantly improved when a biosurfactant was present [177]. Owsianiak et al. [178] demonstrated that biodegradation of oil hydrocarbons was dependent on composition and concentration and the added biosurfactant stabilized the hydrocarbon–water emulsion. Hence, for stimulating biosurfactant producing microorganisms, one must also consider the oil composition (e.g., chain length, aromatic fraction). Also the carbon and nitrogen source of the biosurfactant producing microorganisms determines biosurfactant production and γ decrease [177]. Additionally, oil composition also controls the type of bio-degrading microorganisms inhabiting the oil drop (Table A.2). Alkane degraders may quickly convert alkanes to fatty acids that act as biosurfactants by removing the aliphatic fraction from the oil [164]. Organic ring reduction on the other hand proceeds via another mechanism and the end product is CO_2 if the carboxylated intermediate is not excreted [179, 180].

A number of aerobic and anaerobic hydrocarbon degrading microorganisms were isolated (Table A.2). Aliphatic or aromatic hydrocarbons serve as energy and carbon sources. Alkanes are often incorporated into biomass as cell inclusions [181–186] or membrane fatty acids [114, 187–190]. As such, the converted oil hydrocarbons may be released as polar lipids upon cell lysis and serve as biosurfactants [170]. Some alkanes also serve as substrates for the production of biosurfactants by yeasts and some prokaryotes (reviewed in [164]). Direct degradation products of aliphatic hydrocarbons are their carboxylates. Aliphatic carboxylates gave oil recoveries of 60–80 % [191]. Also the incorporation of aromatic [112, 192, 193] and polyaromatic hydrocarbons [194–196] into biomass has been reported. Aromatic compounds and their derivates are biodegraded via their respective carboxylates [197]. Many benzoate derivates are powerful surfactants [198]. Also polyaromatic carboxylates (e.g., naphthalene carboxylic acids) are powerful pH-dependent surfactants that are often found in crude oil. The production of such intrinsic biosurfactants may be stimulated by the addition of nitrate as electron acceptor [199]. Ultimately, dead cells or cell debris comprising nonionic SAC and polar lipids could act as surfactants or solvents resulting in increased oil droplet mobility [200]. Sequential feeding and starvation (to lyse cells) may be a simple method to produce such a mix of biosurfactants and would not require injection of specific biosurfactant producers.

However, injecting biosurfactant producing microorganisms into an oilfield may be profitable if the process of microbial growth can be controlled. One comparative MEOR study clearly demonstrated the migration of injected biosurfactant producing *Bacillus* strains within the Bebee oilfield, OK, USA [55]. *Bacillus* strains can produce surfactin (Table A.3). This was well studied using *Bacillus subtilis* and other bacilli [166]. Bacilli have the advantage that they can be grown quickly in the lab under oxic conditions but strive as well under anoxic conditions such as *Bacillus mojavensis* JF2. Albeit capable of reducing the γ between the aqueous phase and hexadecane below 1 mN m^{-1}, surfactin was rarely employed in MEOR studies beyond γ measurements [201]. One study compared the efficiency of lipopeptide producing microorganisms in three injection and the respective production wells [55]. The authors used an inoculum of two *Bacillus* strains along with nutrients (glucose, nitrate, and trace elements) for two wells. Sufficient biosurfactant production for oil displacement (up to 90 mg l^{-1}) was observed in the production fluids along with the injected microorganisms of the inoculated wells. The total carbon balance was 107 % thus moderately exceeding the injected amount of nutrients. Also the surface tension of the inoculated wells decreased from 66 and 72 mN m^{-1} to 57 and 56 mN m^{-1}, respectively. It was assumed that this decrease caused the EOR observed.

Reduction of γ by biosurfactants may be further improved by the addition of co-surfactants, such as organic solvents (alcohols, aldehydes and ketones, organic acids, amines, and the like [155, 191]). 2,3-Butanediol was used as co-surfactant along with a partially hydrolyzed polyacrylamid viscosifier in an experiment with *B. mojavensis* JF2 [202, 203]. Up to 45 % residual oil recovery was achieved using this method and γ was reduced to 0.9 mN m^{-1} [204]. Such an additional γ reduction by three orders of magnitude was shown for the glycolipid producing *Rhodococcus* strain H13 [156]. In this study, 0.5 % *n*-pentanol reduced γ against *n*-hexadecane from 1.4 mN m^{-1} to 6.0×10^{-5} mN m^{-1} (Table A.3). The *Rhodococcus* strain was able to synthesize the trehalose lipid surfactant directly from several *n*-alkanes, including *n*-hexadecane. Some publications reported increased oil recovery of about 30–60 % in core flooding experiments, claiming solvent production to be the reason (for a critical review see [144]). These observations must be interpreted carefully because an increase of pressure could have been the underlying mechanism as this is easier to achieve in the small volume of a core, compared with an actual oil reservoir [144].

3.3.4 Salinity, pH, and Temperature and Porous Medium Effects on Biosurfactant Performance

Direct conversion of *n*-alkanes into the biosurfactant surfactin was reported for the *B. subtilis* strain C9 [205]. The γ-decreasing performance of this biosurfactant was strongly dependent on the salinity and the pH of the solution [203, 206, 207]. It was shown that the optimum pH for surfactin was 6.0 and that 5 % NaCl increased the interfacial activity compared with 0.5 % NaCl [207]. Depending on the surfactant/co-surfactant formulation, optimal salinities for oil displacement ranged from 1 to 5 % in another study [191]. Other surfactants seem to have their optimum at alkaline pH, for example, between 8 and 9 [191] or at 10, [177]. In phase mixing experiments it was discovered that an optimum salinity of 7 % was required for best mixing brine and crude oil [206]. The γ value could be reduced to 0.025 mN m^{-1} by a rhamnolipid between toluene and the brine at an optimal salinity of 3 % (Table A.3; [208]). In contrast, an adverse effect of higher salinity on the surface tension reduction by two different *Bacillus* strains and one *Pseudomonas* strain was observed but was the least negative for *Bacillus cereus* [169]. Increasing γ values along with higher salinities were also observed in a study with *Enterobacter cloacae* and *Pseudomonas* sp. and their cell free supernatants [177].

Another factor influencing γ reduction is temperature. Zekri et al. [209] demonstrated an additional γ reduction by three orders of magnitude when the temperature of a biosurfactant producing microbial solution rose above 60 °C. However, this was likely due to the thermophilic nature of the cultures and their higher activity. Interfacial activity depending on temperature was also observed in a study of a biosurfactant producing consortium of *E. cloacae* and *Pseudomonas* sp. [177]. In the range from 20 to 70 °C, the greatest γ reduction was observed at 40 °C. This effect was dependent on cell growth and surfactant production because the cell free surfactant solution was stable over a wide temperature range from 4 to 121 °C. In conclusion, the biosurfactant that is to be produced in the oilfield must be active under the given conditions, such as pH, salinity, and temperature. Only *Bacillus* biosurfactants (surfactin, lichenysin A) have been characterized sufficiently (structure, activity, genetics) to allow a forecast about their activity in the field [76, 207, 210–214].

Another issue is that 0.1 to 1 mg biosurfactant adsorbs effectively to 1 g rock and is lost for oil displacement [144]. However, this can be circumvented by the production of "sacrificial" biosurfactants that block the mineral matrix which may lead to a 25–30 % reduction in adsorption of the more effective surfactants [144]. It is also obvious that the local production of biosurfactants at entrapped oil droplets will reduce these losses and will therefore likely improve the carbon balance. Moreover, biodegradation of biosurfactants by competing microorganisms can result in additional losses [47]. The wettability of the rock surface is crucial for the adsorption of oil and surfactants. This depends on the mineral composition of the rock. Wettability plays a minor role for oil displacement from sandstone formations [144]. In the case of carbonates and clays, the rock wettability is crucial for oil and surfactant adsorption. Changing the wettability of carbonate rocks using biosurfactants, solvents, or polymers (biofilms) seems an interesting option. In contrast, the wettability may become insignificant when oil droplets are coated with a biofilm of hydrocarbon degrading microorganisms [157, 215, 216]. Consequently, it has recently been proposed that microorganisms coating oil droplets increased the oil production in a commercial MEOR application [217].

3.4 Effects of Oil and Water Viscosity

Entrapped oil and surrounding water have different viscosities. This is one of the factors that make these two fluids immiscible and retain the oil in reservoir pores. An increase of the water viscosity or a decrease of the oil viscosity may enhance oil displacement. Microbial products that increase the water viscosity are biofilms and other biopolymers, fatty acids, or long-chain alcohols. On the other side, the viscosity of the oil can be reduced by organic solvents such as short-chain hydrocarbons, alcohols, ketones, organic acids, or gases.

3.4.1 Increasing Water Viscosity

Microbial isolates obtained from the Tinggi and Semangkok oilfields in Malaysia were able to increase the viscosity of their culture medium from 1.01 to 3.8 cP [218]. A field study in the Chinese Fuyu oilfield could also show an increase of the viscosity of the formation water [41]. An increase of the viscosity to 100 cP in the culture medium was described using the polymer producing strain *Clostridium* sp. TU-15A [39]. Such a hundredfold increase in combination with a tenfold γ decrease would increase the capillary number to a value needed for oil displacement [155]. Although this combined approach appears to be promising in theory, laboratory or field studies investigating this strategy are scarce. Nonetheless, the increase of the viscosity of the aqueous phase alone could reduce the volumetric flow and the overall production, thereby increasing the chance to extract more OIP.

[1] The term "biocracking" refers to thermal cracking, usually used in the refining industry to break large hydrocarbon molecules into smaller ones, which is essentially what microorganisms do at much lower temperature.

3.4.2 Examples of Oil Viscosity Decrease

Microbial cracking or biocracking[1] of long-chain paraffins along with an increasing short-chain/long-chain ratio was observed in numerous field and laboratory studies (e.g., Refs. [170, 219–221]). This was usually accompanied by γ and oil viscosity reduction. In a core flooding experiment using samples from the Changqing oilfield in China, low carbon number products such as CO_2, CH_4, C_2H_6, C_3H_8, *i*- and *n*-alkanes up to C_7, and alkenes were produced [170]. The production of hydrocarbons by microorganisms is well known and might be a viable approach to reduce the OIP viscosity (for review see [222]). Ethane production was demonstrated from ethyl-coenzyme M in a sludge experiment [223] or directly from ethanol by *Methanosarcina barkeri* [224]. The biological reduction of C_{2+} chain alcohols is also thermodynamically feasible under reservoir conditions [225]. Also de novo synthesis of (cyclic) alkenes was shown for some prokaryotes [226–228].

The production of gases seems another viable option to reduce oil viscosity. CO_2 was believed in the Fuyu oilfield in China to have contributed to oil swelling as well as viscosity reduction and ultimately to increased oil production [41]. Although during the study the suspected CO_2 production was not determined, it is routinely used in EOR to drive out residual OIP from porespace [145]. CO_2 is soluble in crude oil and known to decrease its viscosity. Also, methane is a common product in biodegraded oilfields and can reduce the oil viscosity as well [23]. At 79 bar and 50 °C, the solubility of crude oil in methane is 13.6 mg l^{-1} but depends on the respective oil composition [229]. The solubility of crude oils in water with 3 % salinity was between 8 and 20 mg l^{-1} and between 50 and 400 μg l^{-1} of the weathered fraction, respectively [230]. The effect of methane on OIP solubility in water has not been studied yet, but an increase seems likely. However, due to the high solubility of crude oil in methane, crude oil viscosity will decrease resulting in higher porespace mobility [231]. Methanogenic hydrocarbon degradation from various aliphatic and aromatic compounds is exergonic under physiological standard conditions (Table 1). Its degradation can be stimulated by inoculation, as shown in a sandstone core experiment [140], or by accelerating β-oxidation of intermediate fatty acids [141]. β-Oxidation rates of intermediate fatty acids appear to be the bottleneck in reduced environments such as oil reservoirs as it requires an efficient electron sink [232], but can be encountered by the addition of electron acceptors [141]. However, nitrate inhibits methane production, but its microbial turnover product N_2 may dissolve in the OIP and decrease its viscosity [233]. For accelerated degradation of aromatic compounds, ring reduction is involved. It has been demonstrated for toluene that an electron accepting anode can accelerate its degradation [234], but it remains unclear if this works for more complex ring structures that are usually observed in residual oil. Although it is possible that microorganisms metabolize long-chain hydrocarbons into shorter hydrocarbons via biomass synthesis, microcosm studies demonstrating biocracking in oilfield samples must demonstrate this in each specific setting.

3.5 Porespace Alteration by Plugging or Rock Dissolution

Generally, the mechanism of microbial rock colonization is poorly understood. However, it is reasonable to assume that greater porespace allows greater substrate fluxes and hence a faster colonization by microorganisms. Thus, reservoir zones of large pore sizes can be colonized and plugged faster than zones of little porosity. Highly permeable reservoir zones of low sweep efficiency are characterized by high flux rates and low OIP concentrations whereas the major part of the OIP may reside in impermeable zones with low fluxes. The concept of selective plugging allows microorganisms to block the highly permeable zones to direct the sweep fluids to less permeable zones with higher oil saturations. Three biological mechanisms can block high permeable zones: (i) the formation biofilms, (ii) the formation of highly viscous organic matter [235], or (iii) the precipitation of inorganic minerals.

Porespace alteration received much attention throughout the last three decades but the relevance of MEOR was demonstrated in several laboratory and field studies [39, 209, 236–240]. The preconditions for successful applications were briefly reviewed by Gray et al. [144]. These are: high travel velocities of the injected fluids as well as injected microorganisms that survive under reservoir conditions and out-compete the indigenous community. Moreover, the produced polymer must not be biodegradable. Ideally, the injected microorganisms do not degrade petroleum to minimize the risk of colonizing the low permeability zones laden with hydrocarbons. This would possibly have the unintended effect of blocking the extractable OIP. Based on Stiles permeability calculations [241], Gray et al. [144] concluded that a porosity of less than 6 % would be suitable for microbial plugging. Plugging reservoir fractures would have better prospects. The advantage of this method is that it could be applied to any depleted oilfield when the injected microorganisms were previously isolated from the same field [242]. For economic reasons, marginal wells should be the primary target [144]. The selective plugging method was successfully applied during multiwell pilot studies, for example, Refs. [242–244], and was recently reviewed in detail by Youssef et al. [7]. In addition to polymers produced by microorganisms, the precipitation of inorganics, so-called mineral plugging, could be a feasible strategy [236]. It is well known that microorganisms are able to accelerate precipitation processes to form solid carbonates [245–248] or sulfides [46, 249].

In addition to plugging, rock dissolution can liberate entrapped OIP. It was demonstrated that certain microorganisms (e.g., *Cryptoendolithic lichens* [250]) are able to alter the surface of sandstone under aerobic conditions by producing oxalic acid and thereby mobilize iron compounds [250–252]. This could increase the porosity of a reservoir if similar compounds are formed under anaerobic conditions. Moreover, the microbial production of acids hydrolyzes carbonates. It was demonstrated in core flooding experiments that carbonate dissolution may indeed liberate oil from carbonate formations [5]. In some carbonate reservoirs this mechanism was believed to be the reason for higher porosity [152, 253–255].

4 The Economy of MEOR

In 1987, the price of one bbl oil was US $16.50 (Fig. 2). By that time, Burchfield and Carroll Jr. [256] calculated a breakeven of costs and yield after four months MEOR treatment, carried out as single well stimulation. During the following 10 years, the oil price remained stable and Portwood [257] estimated average MEOR treatment costs of US $0.50 per bbl and US $2.00 for each incremental bbl. This means that all produced oil would cost an additional US $0.50 per bbl and the costs for each incremental bbl alone would be US $2.00. For example, a well produces 3 bbl day^{-1} and after MEOR treatment of US $2.00 day^{-1} the production inclines to 4 bbl day^{-1}. The total production costs inclined by US $2.00 day^{-1}, which are divided by the total amount of produced bbls. If the production costs were US $16.00 per bbl before MEOR, the costs were US $16.50 after MEOR commenced. At an oil price of US $16.50 in 1987, this was the breakeven. In March 2013, the oil price was US $91.02 for WTI and US $110.40 for Brent (Fig. 2). Considering this and a cumulative inflation of only 50 % between 1995 and 2012, the MEOR process is only a small investment compared with potential gains. Similar conclusions were made for a 7-year MEOR project conducted on the La Ventana oilfield [258]. The authors calculated additional treatment costs of US $1.39 to 2.35 per bbl, costs that were reported from other field tests and conventional EOR methods as well [259]. Based on laboratory experiments, Vadie et al. [260] used 0.12 % KNO_2 and 0.06 % NaH_2PO_4 to stimulate growth of indigenous microorganisms and observed a declining water cut compared to a control well. The same formulation was used in a recent commercial MEOR treatment, claiming an average sustained surplus of oil production of 200 % with treatment costs of US $6.00 per incremental bbl [217]. The use of indigenous microorganisms was also suggested in a critical survey evaluating the economy of MEOR because the likelihood of thriving under reservoir conditions is greater than for laboratory strains [259]. Successful field experiments employing inexpensive nitrate were conducted in Canada, China, and the United States [233]. KNO_3 can be purchased for US $200–500 per ton and NaH_2PO_4 for US $500–800 per ton. It should be approximated in reservoir core experiments how much is needed for successful oil detachment because the addition of too much nitrate bears the risk of OIP degradation. Another study compared biocracking with conventional hot oil treatment (HOT, a classical EOR method where hot water or steam is injected [261]) and an average reduction of the treatment costs from US $4.01 per bbl for HOT to US $2.02 per bbl was achieved [262]. This was in good agreement with estimated costs for MEOR in a mature Argentine oilfield of US $2 for each incremental bbl, based on a field pilot study [263]. In

1998, a pilot MEOR study in the Changqing oilfield in China resulted in an overall profit of US $83,000 treating 27 wells [170].

In the case of biosurfactant production, Gray et al. [144] estimated that 204 tons of biosurfactant are needed to recover 145,000 bbl of incremental oil when the residual OIP was 30 % and the incremental oil recovery was 15 %. In the same report the authors calculated that 146 kg alkanes must be converted to yield 1 bbl incremental crude oil [144]. Glucose and a microbial inoculum may be injected but in this case 204 kg of glucose + microbial cells are necessary to yield one bbl of crude oil. Taking only materials into account, very low treatment costs (US $1.6 per bbl), were reported from an inoculation study using biosurfactant producing *Bacillus* strains in an oilfield in OK, USA [55]. Treatment costs of less than US $1 per bbl were reported from an oilfield in California, USA, again, not considering public co-funding [233]. If the surfactant can be recovered and reinjected, then the costs for the MEOR treatment may further decrease. Considering that cell debris may act as surfactant as well, the alkane to crude oil ratio may considerably decrease.

It must be noted that most economic estimations were made for marginal oilfields often owned by minor producers and producing at the economic limit. Although it was suggested that such oilfields may be ideal for MEOR treatment [144], large oilfields such as the Beatrice and Gullfaks oilfields in the North Sea or Chinese oilfields like Dagang and Chaqing were MEOR treated as well. The effect of MEOR was difficult to assess in a North Sea MEOR study [264]. Still, the North Sea Beatrice field has produced oil for more than 15 years beyond the expected closure, probably due to a combination of MEOR treatment and oilfield upgrading. Economic data from other oilfields are difficult to interpret because success reports are sometimes hardly different from propaganda. Nevertheless, MEOR appears to be successfully applied in numerous cases in Chinese oilfields [265].

5 Conclusions

A recent review of 10 MEOR field applications concluded that the best results were obtained in reservoirs with a high water cut (>75 $\%_{v/v}$), low salinity (10 mol %), and temperature (<55 °C; [266]). This review reported a success of 80 %, however, an extensive recent MEOR survey of 403 laboratory and field studies, including well stimulation, reported 96 % successful projects [7]. In another recent MEOR field study, the authors concluded that no difference in oil recovery is to be expected when MEOR starts with waterflood or at a later stage [144]. The aforementioned surveys were made by MEOR experts who were likely strong supporters of MEOR. Moreover, it is traditionally difficult to publish negative results. Hence, all MEOR studies and reviews have to be interpreted very carefully. Nevertheless, some critical reviews are available [144, 259].

Despite decades of laboratory research and many successful field trials, MEOR is not yet accepted as a routine EOR procedure by the oil industry. There is still a degree of skepticism due to the lack of reliability and the difficulty in predicting the results of MEOR procedures. Moreover, investigators did not agree on a universal MEOR mechanism. Here, we suggest that the production of methane or biosurfactants in situ or offsite in bioreactors might be the most promising MEOR strategies. Generally, too little is known about microbial processes in oilfields. It is therefore necessary to study these processes using classical culturing techniques in combination with established methods targeting DNA or RNA. Stable isotope probing can shed light on key players in such processes. Better and cheaper tools to investigate the activity of microorganisms at oil–water interfaces must be developed. It still remains challenging to control microbial processes in oil and coal deposits. Nevertheless, some promising tools to study complex microbial processes are available and should be used.

Acknowledgments We are indebted to the anonymous reviewer for helpful comments as well as to Matthew Yates of the Pennsylvania State University for correcting the English language of this text.

Appendix

See Tables A.1, A.2 and A.3.

Table A.1 Microbial strains that were isolated from oilfields and their growth conditions

Species	Hydrocarbon substrate	Oxygen utilization	Salinity condition (%)	Growth temperature (°C)	Growth pH	Biofilm forming	Surfactant producing	Reference	Isolation source
Achromobacter xylosoxidans	Aliphatic	Aerobic	(5)	(37)	(7.0)	Unk.	Yes	[267]	Ramoura oilfield, Tunisia
Anaerobaculum thermoterrenum	Unk.	Anaerobic	0–20 (1)	28–60 (55)	5.5–8.6 (7.0–7.6)	Unk.	Unk.	[268]	Redwash oilfield in UT, USA
Bacillus licheniformis	Aliphatic/aromatic	Aerobic/anaerobic	1.1–15	30–55	(7.0)	Unk.	Unk.	[45, 267]	Natural oilfield, Atlantic Ocean, Sercina oilfield, Tunisia
Bacillus thermoleovorans	Aliphatic	Aerobic	0.1–0.5 (0.1)	50–80 (65–70)	5.0–8.0 (6.5–7.0)	Unk.	Unk.	[269]	Minami-aga, Yabase oilfields Japan (95 °C)
Brevibacillus thermoruber	Aliphatic	Aerobic	(1.1)	(55)	(7.0)	Unk.	Yes	[267]	Sercina oilfield, Tunisia
Caldanaerobacter subterraneus	Unk.	Anaerobic	0–3 (0)	40–75 (65)	6.0–8.5 (7.5)	Unk.	Unk.	[270, 271]	Gram positive, Lacq Supérieur oilfield, France
Cyclobacterium lianum	Unk.	Aerobic	0.1–12 (1–4)	15–42 (33)	6.5–9.0 (7.5–8.0)	Unk.	Unk.	[272]	Xijiang oilfield, China
Deferribacter thermophilus	No	Anaerobic	0–6	50–65	5–8	Unk.	Unk.	[273]	Beatrice oilfield, North Sea
Desulfacinum infernum	No	Anaerobic	0–50	40–65 (60)	6.6–8.4	Unk.	Unk.	[274]	North Sea oilfield
Desulfacinum subterraneum	Unk.	Anaerobic	0–50	45–65 (60)	6.5–8.4	Unk.	Unk.	[275]	Stratal brine of an oilfield
Desulfobacter vibrioformis	Unk.	Anaerobic	1–5	5–38 (33)	6.8–7.0	Unk.	Unk.	[276]	North Sea oilfield
Desulfobacterium cetonicum	Aromatic	Anaerobic	1–5 (1–2.8)	4–37 (30–32)	6.0–7.9 (7.2–7.5)	Unk.	Unk.	[13, 277]	Absheron Peninsula oilfield
Desulfobulbus rhabdoformis	Unk.	Anaerobic	1.5–2.0	10–40 (31)	6.8–7.2	No	Unk.	[278]	North Sea oilfield
Desulfococcus oleovorans Hxd3	Aliphatic	Anaerobic	2.8	20–30	7	Unk.	Unk.	[279]	German Oilfield near Hamburg
Desulfoglaeba alkanexedens	Aliphatic	Anaerobic	0–5.5	17–50 (31–37)	4.5–8.2 (6.5–7.2)	Yes	Unk.	[280]	Oilfield in OK, USA
Desulfomicrobium apsheronum	Unk.	Anaerobic	0–8 (1)	15–40 (25–30)	7.0–7.2	Unk.	Unk.	[14]	Absheron Peninsula, oilfield
Desulfotomaculum halophilum	Unk.	Anaerobic	1–14 (4–6)	30–40 (35)	6.9–8.0 (7.3)	Unk.	Unk.	[281]	Paris Basin oilwell
Desulfotomaculum thermocisternum	No	Anaerobic	0.0001–4.6 (0.3–1.3)	41–75 (62)	6.2–8.9 (6.9)	Unk.	Unk.	[282]	Syntrophy with methanogen, North Sea oil reservoir
Desulfovermiculus halophilus	Unk.	Anaerobic	3.0–2.3	25–47	6.0–8.5	Unk.	Unk.	[283]	Stratal brine of an oilfield
Desulfovibrio alaskensis	Unk.	Anaerobic	0–10 (2.5)	10–45 (37)	6.5–8.5 (7.0)	Unk.	Unk.	[284]	Purdu Bay, AK, USA
Desulfovibrio bastinii	Unk.	Anaerobic	1–12 (4)	20–50 (35–40)	5.2–7.4 (5.8–6.2)	Unk.	Unk.	[285]	Emeraude Oilfield, Congo
Desulfovibrio capillatus	Unk.	Anaerobic	0.5–10 (1.5–3)	20–50 (40)	6.5–8.8 (7.4)	Unk.	Unk.	[286]	Gulf of Mexico

(continued)

Table A.1 (continued)

Species	Hydrocarbon substrate	Oxygen utilization	Salinity condition (%)	Growth temperature (°C)	Growth pH	Biofilm forming	Surfactant producing	Reference	Isolation source
Desulfovibrio gabonensis	Unk.	Anaerobic	1–17 (5–6)	15–40 (30)	6.4–8.2 (6.9–7.3)	Unk.	Unk.	[287]	African offshore pipeline
Desulfovibrio gracilis	Unk.	Anaerobic	2–12 (5–6)	20–40 (37–40)	5.4–8.4 (6.8–7.2)	Unk.	Unk.	[285]	Emeraude Oilfield, Congo
Desulfovibrio longus	Unk.	Anaerobic	0–8 (1–2)	10–40 (35)	6.5–8.5 (7.4)	Unk.	Unk.	[288]	Paris Basin oil well
Desulfovibrio vietnamensis	Unk.	Anaerobic	0–10 (5)	12–45 (37)	5.0–10.0 (7.5)	Unk.	Unk.	[289]	Vietnamese oilfield
Dethiosulfovibrio peptidovorans	Unk.	Anaerobic	(3)	20–45 (42)	(7.0)	Unk.	Unk.	[290]	Emeraude oilfield, Congo
Dietzia sp.	Aliphatic/ aromatic	Aerobic	(1.1–1.5)	(30)	(7.2–7.6)	Unk.	Unk.	[291]	Daqing oilfield, China
Fusobacter paucivorans	Unk.	Anaerobic	0–10 (0–3)	20–45 (37)	5.7–8.0 (7.3)	Unk.	Unk.	[292]	Emeraude oilfield, Congo
Garciella nitratireducens	Unk.	Anaerobic	0–10 (1)	25–60 (55)	5.5–9.0 (7.5)	Unk.	Unk.	[293]	SAMII oilfield, Gulf of Mexico
Geoalkalibacter subterraneus	No	Anaerobic	0.1–10	30–50	6–9	Unk.	Unk.	[294]	Redwash oilfield in UT, USA
Geobacillus jurassicus	Aliphatic	Aerobic	0–5 (0.5)	45–65 (58–60)	6.4–7.8 (7.0–7.2)	Unk.	Unk.	[295]	Dagang oilfield, China
Geobacillus kaustophilus	Aliphatic	Aerobic	Unk.	40–75 (n/d)	6.2–7.5	Unk.	Unk.	[296]	unknown oilfield
Geobacillus lituanicus	Unk.	Aerobic	(0–0.5)	55–70 (55–60)	(6.5)	Unk.	Unk.	[297]	Girkaliai oilfield, Lithuania
Geobacillus pallidus	Aromatic/ aliphatic	Aerobic	(1.1–1.6)	(55)	(7.0)	Unk.	Yes	[267, 298]	Litayem oilfield, Tunesia
Geobacillus sp. SH-1	Aromatic/ aliphatic	Aerobic	(1–3)	45–80 (70)	6.0–8.0 (7.0)	Unk.	Unk.	[299]	Shengli oilfield, China
Geobacillus subterraneus	Aliphatic	Aerobic	0–5	45–70 (n/d)	6.0–7.8	Unk.	Unk.	[296]	Samotlor oilfeld, Western Siberia, Russia; Liaohe oilfield, China
Geobacillus thermocatenulatus	Unk.	Aerobic	0–1.5	42–69 (n/d)	6.5–8.5	Unk.	Unk.	[296]	Gas field
Geobacillus uzenensis strains	Aliphatic	Aerobic	0–4	45–65 (n/d)	6.2–7.8	Unk.	Unk.	[296]	Uzen oilfeld, Kazakhstan
Gordonia paraffinivorans	Unk.	Aerobic	0.5–7	20–45 (30–37)	5.5–9.5	Unk.	Unk.	[300]	Gram positive, Daqing oilfield, China
Haloanaerobium congolense	Unk.	Anaerobic	4–24 (10)	20–45 (42)	6.3–8.5 (7.0)	Unk.	Unk.	[301]	Congolese oilfield
Halomonas lutea	Aliphatic	Aerobic	(10)	(37)	(7.0)	Unk.	Yes	[267]	Sercina oilfield, Tunesia
Kosmotoga olearia	Unk.	Anaerobic	1–6 (2.5–3.0)	20–80 (65)	5.5–8.0 (6.8)	No	Unk.	[302]	Troll oilfield North Sea
Lysinibacillus fusiformis	Aliphatic	Aerobic	(5)	(37)	(7.0)	Unk.	Yes	[267]	Ramoura oilfield, Tunesia
Mahella australiensis	Unk.	Anaerobic	<=4 (0.1)	30–30 (50)	5.5–8.8 (7.5)	Unk.	Unk.	[303]	Riverslea cilfield, Australia
Methanobacterium sp.	Unk.	Anaerobic	0.1–3.1 (n/d)	37–45 (n/d)	6.8–7.2 (7.0)	Unk.	Unk.	[53]	Bondyuzhskoe oilfield, Tatarstan, Russia

(continued)

Table A.1 (continued)

Species	Hydrocarbon substrate	Oxygen utilization	Salinity condition (%)	Growth temperature (°C)	Growth pH	Biofilm forming	Surfactant producing	Reference	Isolation source
Methanocalculus halotolerans	Unk.	Anaerobic	0–12.5 (5)	24–50 (38)	7.0–8.0 (7.6)	Unk.	Unk.	[304]	Methanogen, Scheibenhardt oilfield, France
Methanococcus thermolithotrophicus	Unk.	Anaerobic	0.6–9.4 (1.5–2.3)	17–62 (60)	4.9–9.8 (5.1–5.9)	Unk.	Unk.	[305]	Methanogen, North Sea oilfield
Methermicoccus shengliensis	Unk.	Anaerobic	1.2–6.4 (1.8–2.9)	50–70 (65)	5.5–8.0 (6.0–6.4)	Unk.	Unk.	[306]	Methanogen, Shengli oilfield, China
Modicisalibacter tunisiensis	Unk.	Aerobic	1–25 (10)	4–45 (37)	5–10 (7.2)	Unk.	Unk.	[307]	Sidi Litayem oilfield, Tunisia
Nocardioides oleivorans	Yes	Aerobic	(2)	(25–30)	(7.3)	Unk.	Unk.	[308]	Gram positive, Oerel oilfield, Germany
Oceanotoga teriensis	Unk.	Anaerobic	0–12 (4–4.5)	25–70 (55–58)	5.5–9.0 (7.3–7.8)	Unk.	Unk.	[309]	Bombay High oilfield, India
Petrimonas sulfuriphila	Unk.	Anaerobic	0–4 (0)	15–40 (37–40)	(7.2)	Unk.	Unk.	[310]	Canadian oilfield
Petrobacter succinatimandens	Unk.	Anaerobic	<=3 (0.5)	35–60 (55)	5.5–8.0 (6.9)	Unk.	Unk.	[311]	Riverslea oilfield, Australia
Petrotoga halophila	Unk.	Anaerobic	0.5–9 (4–6)	45–65 (60)	5.6–7.8 (6.7–7.2)	Unk.	Unk.	[312]	Tchibouella oilfield, Congo
Petrotoga sp.	Aliphatic	Anaerobic	<3 (n/d)	50–70 (n/d)	Unk.	Unk.	Unk.	[313]	Yabase oilfield, Japan
Petrotoga mexicana	Unk.	Anaerobic	1–20 (3)	30–65 (55)	5.9–8.5 (6.6)	Unk.	Unk.	[314]	Mexican oilfield, Gulf of Mexico
Petrotoga mobilis	Unk.	Anaerobic	0.5–9.0 (3–4)	40–65 (58–60)	5.5–8.5 (6.5–7.0)	Unk.	Unk.	[315]	North Sea oilfield
Pseudomonas aeruginosa	Unk.	Anaerobic	5–15 (5)	37–70 (37)	6.3–7.3 (7.0)	Unk.	Yes	[267, 316]	Egyptian oilfield, Gulf of Suez, Ramoura oilfield, Tunisia
Pseudomonas stutzerii	Aliphatic	Aerobic	(1.1)	(37)	(7.0)	Unk.	Yes	[267]	Sercina oilfield, Tunesia
Rhodococcus ruber	Aliphatic	Aerobic/ anaerobic	(0.7–1.0)	(37)	(7.0–7.5)	Unk.	Yes	[317]	Daqing oilfield, China
Spirochaeta smaragdinae	Unk.	Anaerobic	1–10	20–40	5.5–8.0	No	Unk.	[318]	Congolese oilfield
Thermoanaerobacter brockii subsp. *lactiethylicus*	Unk.	Anaerobic	1–2 (2)	40–75 (55–60)	(7.3)	Unk.	Unk.	[319]	French oilfield
Thermococcoides shengliensis	Unk.	Anaerobic	0–4 (1.5)	45–75 (65)	6.0–8.0 (7.0)	Unk.	Unk.	[320]	Shengli oilfield, China
Thermodesulfobacterium norvegicus	Unk.	Anaerobic	0–5.6 (1.6)	44–74 (60)	6.1–7.7 (6.9)	Unk.	Unk.	[321]	North Sea oilfield
Thermotoga elfii	Unk.	Anaerobic	0–2.8 (1)	50–72 (66)	5.0– <9.1 (7.5)	No	Unk.	[322]	African oilfield

(continued)

Table A.1 (continued)

Species	Hydrocarbon substrate	Oxygen utilization	Salinity condition (%)	Growth temperature (°C)	Growth pH	Biofilm forming	Surfactant producing	Reference	Isolation source
Thermovirga lienii	Unk.	Anaerobic	5–8 (2–3)	37–68 (58)	6.2–8.0 (6.5–7.0)	Unk.	Unk.	[323]	Troll oilfield North Sea
Wenxinia marina	Unk.	Aerobic	0.5–9 (1–4)	15–42 (34–38)	6.5–8.5 (7.5–8.0)	Unk.	Unk.	[324]	Xijiang oilfield, China

Numbers in brackets give optimal growth conditions. Values without brackets indicate the growth range. When the publication does not distinguish between growth range and optimal growth, it was assumed that the given number represents optimal conditions and was placed in brackets

Unk unknown, *n/d* not determined

Table A.2 Microbial isolates that are able to degrade hydrocarbons

Species	Hydrocarbon substrate	Oxygen utilization	Salinity condition (%)	Growth temperature (°C)	Growth pH	Biofilm forming	Surfactant producing	Reference	Isolation source/remarks
Acidocella sp.	Aromatic	Aerobic	(0.1)	Unk.	(4–6)	Unk.	Unk.	[325]	Acidic soil surrounding outdoor coal storage pile in northern Indiana
Acidovorax sp.	Aromatic	Aerobic	(0.4)	(25)	(6.9)	Unk.	Unk.	[326]	Municipal wastewater plant, Gießen, Germany
Acinetobacter baumanii	Fuel oil	Aerobic	(0.4)	(24)	Unk.	Unk.	Unk.	[327]	Unpolluted temperate agricultural soil, Montmirail, Marne, France
Acinetobacter lwoffi	Aliphatic	Aerobic	(3.7)	(25–30)	(7.1)	Unk.	Unk.	[328]	Activated sludge, Canterbury Sewage Works, Kent, UK
Acinetobacter venetianus	Aliphatic	Aerobic	(3.2)	(30)	(7.0)	Unk.	RAG-1	[172, 329]	Seawater from a local beach Tel Baruch, Tel Aviv, Israel; reclassified
Acinetobacterium calcoaceticum	Aliphatic	Aerobic	(0.03)	(28)	(7.0–7.2)	Yes	Yes	[330, 331]	Altered heating oil from storage tank
Actinomyces chromogenes	Aliphatic	Aerobic	(0.03)	(27)	(7.0–7.2)	Unk.	Unk.	[332]	Various sources
Aeromonas sp.	Diesel oil	Aerobic	(0.85)	(30)	Unk.	Unk.	Unk.	[333]	Rainforest soil, University of Port Harcourt, Nigeria
Agrobacterium sp.	Aromatic	Aerobic	(0.03)	(28)	(7.0–7.2)	Unk.	Unk.	[334]	Petroleum contaminated soil near petroleum storage units Rio de Janeiro, Brazil
Alcaligenes denitrificans	Aromatic/ aliphatic	Aerobic	(1)	(23)	Unk.	Unk.	Unk.	[335]	Well water and core material contaminated with unleaded gasoline from a shallow costal aquifer, US Naval Weapons Station, Seal Beach, CA, USA; 220 strains isolated

(continued)

Table A.2 (continued)

Species	Hydrocarbon substrate	Oxygen utilization	Salinity condition (%)	Growth temperature (°C)	Growth pH	Biofilm forming	Surfactant producing	Reference	Isolation source/remarks
Alcaligenes sp.	Aromatic	Aerobic	(0.03)	(28)	(7.0–7.2)	Unk.	Unk.	[334]	Petroleum contaminated soil near petroleum storage units Rio de Janeiro, Brazil
Alcaligenes xylosoxidans	Aliphatic	Aerobic	(0.03)	(30)	(7.0–7.2)	Unk.	Unk.	[336]	Shoot interior of plants potted with crude oil contaminated soil of oil pumping sites, Zisterdorf, Lower Austria
Alcanivorax borkumensis	Aliphatic	Aerobic	1.0–12.5	4–35	(7.5)	Yes	Glycolipids	[337]	Seawater/sediment near Borkum Island (North Sea, Germany)
Alcanivorax dieselolei	Aliphatic	Aerobic	1–15 (3–7.5)	15–45 (28)	(7.5)	Unk.	Unk.	[338]	Strains: oil contaminated surface water Bohai Sea near Shengli oilfield/ deep sea sediments from east Pacific Ocean (Pacific nodule region)
Alcanivorax jadensis	Aliphatic	Aerobic	0.5–15	(25)	(7.2)	Unk.	Unk.	[339, 340]	Intertidal sediment Jadebusen Bay, North Sea coast, Germany
Alcanivorax venustensis	Aliphatic	Aerobic	≤15	4–40	(7.4)	Yes	Unk.	[340]	Strains: seawater near coast of Alicante, Mediterranean Sea, Spain and seawater near coast Santa Pola, Alicante, Spain
Alicycliphilus denitrificans	Aromatic	Aerobic/ anaerobic	(2.6)	15–40 (30–37)	6.6–9.0 (7.3)	Unk.	Unk.	[341]	Mixed sample from wastewater treatment plant and benzene contaminated soil

(continued)

Table A.2 (continued)

Species	Hydrocarbon substrate	Oxygen utilization	Salinity condition (%)	Growth temperature (°C)	Growth pH	Biofilm forming	Surfactant producing	Reference	Isolation source/remarks
Alkanindiges illinoisensis	Aliphatic	Aerobic	(1–10)	(25)	Unk.	Unk.	Unk.	[181]	Crude oil contaminated soil, IL, USA
Alteromonas distincta	Aromatic	Aerobic	Seawater	(30)	(7.8)	Unk.	Unk.	[342]	Seawater 1 m depth Heita Bay, Japan; 15 synthetic surfactants without effect
Archaeoglobus fulgidus	Aliphatic (alkenes)	Anaerobic	(2.1)	(70)	(7.0)	Unk.	Unk.	[343]	Terrestrial heated sea floor at Vulcano, Italy
Aromatoleum sp.	Aromatic/ aliphatic	Anaerobic	(0.2)	(28)	(7.2)	Unk.	Unk.	[197, 344]	Homogenized mixture mud samples from ditches and Weser river in Bremen, Germany
Arthrobacter oxydans	Aliphatic	Aerobic	(0.03)	(30)	(7.0–7.2)	Unk.	Unk.	[336]	Plant rhizosphere potted with crude oil contaminated soil of oil pumping sites, Zisterdorf, Lower Austria
Arthrobacter sp.	Fuel oil	Aerobic	(0.4)	(24)	(3–7)	Unk.	Unk.	[325, 327]	Oil cuttings treated oristine temperate agricultural soil Montmirail, Marne, France and soil surrounding an outdoor coal storage pile in northern IN, USA
Azoarcus EbN1	Aromatic	Anaerobic	(0.2)	(28)	(6.4–8.1)	Unk.	Unk.	[197, 345]	Homogenized mixture mud samples from ditches and Weser river in Bremen, Germany

(continued)

Table A.2 (continued)

Species	Hydrocarbon substrate	Oxygen utilization	Salinity condition (%)	Growth temperature (°C)	Growth pH	Biofilm forming	Surfactant producing	Reference	Isolation source/remarks
Azoarcus toluclasticus	Aromatic	Aerobic/anaerobic	Unk.	(30)	(7.4)	Unk.	Unk.	[346, 347]	Water from experimental Moffet Field aquifer amended with phenol, toluene, chlorinated aliphatic hydrocarbons, Moffett Naval Air Station, Mountain View, CA, USA
Azoarcus tolulyticus	Aromatic	Aerobic/anaerobic	Unk.	(30)	(7.4)	Unk.	Unk.	[348]	Petroleum contaminated freshwater aquifer sediment, 24–25 m deep, northern MI, USA
Azoarcus toluvorans	Aromatic	Aerobic/anaerobic	Unk.	(30)	(7.4)	Unk.	Unk.	[346]	Pristine muck soil, Clinton County, MI, USA and soil from industrial area, Jabaquara, Sao Paulo State, Brazil; (former *A. tolulyticus* [348])
Azotobacter chroococcum	Aliphatic	Aerobic	(2)	(28–30)	Unk.	Unk.	Unk.	[349]	Obtained from Russian Research Institute
Bacillus macroides	Aliphatic	Aerobic	(0.5)	(30)	(7.0–7.2)	Unk.	Unk.	[336]	Plant rhizosphere potted with crude oil contaminated soil of oil pumping sites, Zisterdorf, Lower Austria
Bacillus subtilis	Aliphatic	Aerobic	(0.1)	(30–37)	(6.8)	Unk.	Surfactin	[205, 210]	Soil, details in Ref. [73]
Bacillus subtilis	Aromatic/aliphatic	Aerobic	(0.1)	(27–29)	(7.2)	Unk.	Rhamnolipid	[350]	Hydrocarbon contaminated industrial waste water

(continued)

Table A.2 (continued)

Species	Hydrocarbon substrate	Oxygen utilization	Salinity condition (%)	Growth temperature (°C)	Growth pH	Biofilm forming	Surfactant producing	Reference	Isolation source/remarks
Beijerinckia sp.	Aromatic	Aerobic	(0.1)	(29)	(7.2)	Unk.	Unk.	[351]	Polluted stream Gibson et al. 1993
Blastochloris sulfoviridis	Aromatic	Anaerobic	(1.1)	12–36 (34)	6.5–8.3 (7.4)	Unk.	Unk.	[136]	Activated sludge from sewage plant (Seehausen, Bremen, Germany); phototroph
Brevibacterium erythrogenes	Aliphatic	Aerobic	Unk.	(26)	Unk.	Unk.	Unk.	[352, 353]	Coastal seawater off Sandy Hook, NJ, USA
Brevundimonas vesicularis	Fuel oil	Aerobic	(0.4)	(24)	Unk.	Unk.	Unk.	[327]	Pristine temperate agricultural soil Montmirail, Marne, France treated with oil cuttings
Burkholderia cepacia	Aromatic/ aliphatic	Aerobic	(0.1)	30–42 (30)	(6.8)	Unk.	Unk.	[354, 355]	PAH contaminated soil from an abandoned factory near Port Melbourne, Victoria, Australia and water from holding pond of an industrial waste treatment facility, Naval Air Station, Pensacola, FL, USA
Clavibacter sp.	Aromatic	Aerobic	(0.1)	Unk.	(5–7)	Unk.	Unk.	[325]	Acidic soil surrounding outdoor coal storage pile in northern IN, USA
Clostridium thermosaccharolyticum	Aliphatic	Anaerobic	(0.1)	(56)	(7)	Unk.	Solvents	[356]	Converts paraffin to butanol
Comamonas terrigena	Aromatic	Aerobic	(0.4)	(25)	(6.9)	Unk.	Unk.	[326]	Municipal wastewater plant, Gießen, Germany

(continued)

Table A.2 (continued)

Species	Hydrocarbon substrate	Oxygen utilization	Salinity condition (%)	Growth temperature (°C)	Growth pH	Biofilm forming	Surfactant producing	Reference	Isolation source/remarks
Comamonas testosteroni	Aromatic	Aerobic	(0.4)	(25)	(6.9)	Unk.	Unk.	[326]	Municipal wastewater plant, Gießen,Germany
Corynebacterium hydrocarboclastus	Aliphatic	Aerobic	(1.4)	(25)	(6.4–6.5)	Yes	Polymer	[357]	Refinery effluent wastes
Corynebacterium kutscheri	Aromatic/ crude oil	Aerobic	(0.9)	(30)	(7.0)	Unk.	Unk.	[358]	Soil from Isolo Industrial Estate (abandoned industries of textile, battery, and pharmaceuticals), Lagos, Nigeria
Corynebacterium sp.	Fuel oil	Aerobic	(0.4)	(24)	Unk.	Unk.	Unk.	[327]	Treated and nontreated with oil cuttings unpolluted temperate agricultural soil Montmirail (Marne, France)
Cycloclasticus oligotrophus	Aromatic	Aerobic	Seawater	(30)	(7.8)	Unk.	Unk.	[342]	Deep seawater column near the center of water column in Resurrection Bay, Alaska; 15 synthetic surfactants without effect
Cycloclasticus pugetii	Aromatic	Aerobic	1–7 (3.4)	4–28 (n/d)	6.5–9.5 (7.6)	Unk.	Unk.	[359]	Surface sediment Sinclair Inlet, Puget Sound, Bremerton, WA, USA; Gulf of Mexico sediments near Galveston
Cycloclasticus spirillensus	Aromatic	Aerobic	(3.4)	(23)	(7.6)	Unk.	Unk.	[360]	Burrow wall sediments of benthic macrofauna (mollusc) at the intertidal zone of Lowes Cove, ME

(continued)

Table A.2 (continued)

Species	Hydrocarbon substrate	Oxygen utilization	Salinity condition (%)	Growth temperature (°C)	Growth pH	Biofilm forming	Surfactant producing	Reference	Isolation source/remarks
Dechloromonas aromatica	Aromatic	Anaerobic	0.5	30–35	6.8	Unk.	Unk.	[361, 362]	Sediments collected from the Potomac River, MD, USA
Desulfatibacillum aliphaticivorans	Aliphatic	Anaerobic	0.6–4.5	15–40	6.6–7.8	Unk.	Unk.	[363]	Hydrocarbon-polluted (petroleum refinery spills over a period of 15 years) marine sediment of Canal Vieil cove (Gulf of Fos, France)
Desulfatibacillum alkenivorans	Aliphatic	Anaerobic	0.5–5.0	22–38	6.2–8.0	Unk.	Unk.	[364]	Oil-polluted sediments of a sewage plant of Fos Harbour (France)
Desulfatiferula olefinivorans	Aliphatic	Anaerobic	0–5.0	16–38	6.6–8.3	Unk.	Unk.	[365]	Brackish sediment of a wastewater decantation facility of an oil refinery (Berre lagoon, France)
Desulfitobacterium aromaticivorans	Aromatic	Anaerobic	(0.7)	16–42 (30)	6.5–7.5 (6.6–7.0)	Unk.	Unk.	[366]	Soil of a former coal-gasification site in Gliwice, Poland
Desulfobacula toluolica	Aromatic	Anaerobic	(2.7)	(28)	(7.0–7.1)	Unk.	Unk.	[367]	Anoxic, sulfide-rich marine sediment samples from Eel Pond, a seawater pond in Woods Hole, MA, USA
Desulfosarcina sp. oXyS1	Aromatic	Anaerobic	(2.8)	15–35 (32)	6.2–7.9 (7.5)	Unk.	Unk.	[67, 277]	Water phase (seawater) of a North Sea oil tank in Wilhelmshaven, Germany; strain between *Desulfobacterium* and *Desulfosarcina*

(continued)

Table A.2 (continued)

Species	Hydrocarbon substrate	Oxygen utilization	Salinity condition (%)	Growth temperature (°C)	Growth pH	Biofilm forming	Surfactant producing	Reference	Isolation source/remarks
Desulfosporosinus youngiae	Aromatic	Anaerobic	0.0–3.0	8–39 (32–35)	5.7–8.2 (7.0–7.3)	Unk.	Unk.		Constructed wetland treating acid mine drainage
Desulfothermus naphthae	Aliphatic	Anaerobic	(2.1)	(55–56)	(6.8)	Unk.	Unk.	[368, 369]	Deep marine sediment from Guaymas Basin, 2,000 m depth
Desulfotignum toluenicum	Aromatic	Anaerobic	0.5–5.5 (1.5)	16–35 (34)	6.5–9.0 (7.2)	Unk.	Unk.	[370]	Oil reservoir model column
Desulfotomaculum sp.	Aromatic	Anaerobic	(0.7)	20–36 (25–30)	6.5–7.8 (7.4–7.6)	Yes	Unk.	[371]	Contaminated with mono- and polycyclic hydrocarbon sediment of a drilling core taken at a former gasworks plant near Stuttgart, Germany
Desulfovibrio desulfuricans	Aliphatic	Anaerobic	2	25–30	7	Unk.	Unk.	[134]	Water from artesian wells (depth of 1,280 m, Carrizo Formation) from oil-bearing aquifer southern TX, USA; strain lost
Dietzia psychralcaliphila	Aliphatic	Aerobic	0–10	5–30	7–10	Unk.	Unk.	[372]	Drain pool of a fish-egg-processing plant, water (6 °C, pH 7)
Enterobacter ludwigii	Diesel oil	Aerobic	(0.03)	(30)	(7.0–7.2)	Unk.	Unk.	[373]	Rhizosphere and endosphere of Italian ryegrass and birdsfoot trefoil
Exiguobacterium aurantiacum	Aliphatic	Aerobic	0.2–3.5 (0.2)	(30)	Unk.	Unk.	No	[374]	Contaminated soil

(continued)

Table A.2 (continued)

Species	Hydrocarbon substrate	Oxygen utilization	Salinity condition (%)	Growth temperature (°C)	Growth pH	Biofilm forming	Surfactant producing	Reference	Isolation source/remarks
Flavobacterium balustinum	Aromatic	Aerobic	(1.2)	(4– 40)	Unk.	Unk.	Unk.	[375]	Activated sludge from Institute for Agrobiotechnology (IFA), Tulln, Austria
Flavobacterium sp.	Aromatic/ aliphatic	Aerobic	(0.1)	(29)	(7.2)	Unk.	Unk.	[351, 352]	Brackish water Arthur Kill at Sewaren, NJ, USA and sewage samples from settling tanks Ithaca and Marathon, NY, USA
Gallaecimonas pentaromativorans	Aromatic	Aerobic	0–8 (1)	20–37 (30)	4–10 (5–8)	Unk.	Unk.	[376]	Intertidal sediment of Corcubion Ria in Cee (Prestige oil spill 2002), A Coruna, Spain
Garciaella petrolearia	Aromatic	Aerobic	(0.3)	(50)	Unk.	Unk.	Unk.	[377]	Sea buried oil pipeline Mumbai Uran trunk line (MUT), Bombay, offshore India west coast
Geobacillus stearothermophilus	Aliphatic	Aerobic	0–5	37–65 (n/d)	6.0–8.0	Unk.	Unk.	[296]	Deteriorated canned food (DSMZ information for DSM 22 type strain)
Geobacillus thermoleovorans	Aliphatic	Aerobic	0–4	35–78 (n/d)	6.2–7.8	Unk.	Unk.	[296]	Soil, near hot water effluent, Bethlehem, USA
Geobacter grbiciae	Aromatic	Anaerobic	(0.5)	(30–35)	(6.8)	Unk.	Unk.	[378]	Freshwater sediment of the Potomac River estuary in VA, USA

(continued)

Table A.2 (continued)

Species	Hydrocarbon substrate	Oxygen utilization	Salinity condition (%)	Growth temperature (°C)	Growth pH	Biofilm forming	Surfactant producing	Reference	Isolation source/remarks
Geobacter metallireducens	Aromatic	Anaerobic	(0.5)	(30–35)	(6.8)	Yes	Unk.	[379]	Tidal freshwater river sediment (Eckman dredge) Potomac River, MD, USA
Geobacter toluenoxydans	Aromatic	Anaerobic	(0.7)	16–42 (n/d)	6.5–8.0	Unk.	Unk.	[366]	Well sediment from a tar-oil-contaminated site near Stuttgart, Germany
Georgfuchsia toluolica	Aromatic	Anaerobic	(0.5)	(25–30)	(7.3)	Unk.	Unk.	[380]	BTEX polluted soil, Boxtel, The Netherlands; Fe, Mn, nitrate reducer
Gordonia alkanivorans	Aliphatic	Aerobic	2–10 (6)	13–40 (28)	Unk.	Unk.	Unk.	[381]	Tar contaminated factory soil, Rositz, Thuringia, Germany
Gordonia rubropertincta	Aliphatic	Aerobic	(0.9)	(30)	Unk.	Unk.	Unk.	[382]	Hydrocarbon-contaminated soils, north of Germany
Gordonia sp.	Aromatic	Aerobic	(0.4)	(25)	(6.9)	Unk.	Unk.	[383]	Oil and PAH contaminated soil sites, Hamburg, Germany
Gordonia terrae	Aliphatic	Aerobic	(0.9)	(30)	Unk.	Unk.	Unk.	[382]	Hydrocarbon-contaminated soils, north of Germany
Haloferax alexandrinus	Aromatic	Aerobic	(24)	(40)	(7.2)	Unk.	Unk.	[384]	Hypersaline marshes Uyuni salt flats in Bolivia, crystallizer ponds, Chile and Cabo Rojo (Puerto Rico), and sabkhas (salt flats) Persian Gulf (Saudi Arabia), Dead Sea (Israel and Jordan); Haloarchaeon, PAH

(continued)

Table A.2 (continued)

Species	Hydrocarbon substrate	Oxygen utilization	Salinity condition (%)	Growth temperature (°C)	Growth pH	Biofilm forming	Surfactant producing	Reference	Isolation source/remarks
Haloferax sulfurifontis	Aromatic	Aerobic	(24)	(40)	(7.2)	Unk.	Unk.	[384]	Hypersaline marshes Uyuni salt flats in Bolivia, crystallizer ponds, Chile and Cabo Rojo (Puerto Rico), and sabkhas (salt flats) Persian Gulf (Saudi Arabia), Dead Sea (Israel and Jordan); Haloarchaeon, PAH
Halomonas shengliensis	Crude oil	Aerobic	0–15 (5–15)	10–43 (30)	8.0–9.0 (8.5)	Unk.	Unk.	[385]	Crude oil contaminated saline soil from Shengli oilfield, Shandong Province, China
Klebsiella sp.	Diesel oil	Aerobic	(0.85)	(30)	Unk.	Unk.	Unk.	[333]	Rainforest soil, University of Port Harcourt, Nigeria
Leclercia adecarboxylata	Aromatic	Aerobic	Unk.	(30)	Unk.	Unk.	Unk.	[386]	Subsurface soil of oily-sludge storage pit, Digboi refinery, northeastern India (27°15′ N, 98°1′ E), 100 years contaminated.
Lutibacterium anuloederans	Aromatic	Aerobic	(3.4)	(23)	(7.6)	Unk.	Unk.	[360]	Burrow wall sediments of benthic macrofauna (polychaete) intertidal Lowes Cove, ME, USA
Marinobacter hydrocarbonoclasticus SP17	Aliphatic	Aerobic/ anaerobic	4.7–20.4	10–45 (32)	6.0–9.5	Yes	Biofilm	[157, 387]	Polluted sediments, petroleum refinery outlet, Gulf of Fos, France, 50 km north of Marseille

(continued)

Table A.2 (continued)

Species	Hydrocarbon substrate	Oxygen utilization	Salinity condition (%)	Growth temperature (°C)	Growth pH	Biofilm forming	Surfactant producing	Reference	Isolation source/remarks
Microbacterium hydrocarbonoxydans	Crude oil	Aerobic	2–4 (n.d.)	30–37 (n.d.)	7.3 (n.d.)	Unk.	Unk.	[388]	Oil contaminated soil, Germany
Microbacterium lacus	Aliphatic	Aerobic	(0.03)	(30)	(7.0–7.2)	Unk.	Unk.	[336]	Plant rhizosphere potted with crude oil contaminated soil of oil pumping sites, Zisterdorf, Lower Austria
Microbacterium oleivorans	Crude oil	Aerobic	2–4 (n.d.)	30–37 (n.d.)	7.3 (n.d.)	Unk.	Unk.	[388]	Oil storage cavern 126 near Etzel, Germany
Micrococcus roseus	Crude oil	Aerobic	(0.5)	(30)	Unk.	Unk.	Yes	[389]	Estuarine water samples of the University of Lagos lagoon front, Nigeria
Micrococcus sp.	Aliphatic	Aerobic	(1)	(23)	Unk.	Unk.	Unk.	[335]	Well water and core material contaminated with unleaded gasoline from a shallow costal aquifer, US Naval Weapons Station, Seal Beach, CA, USA; 220 strains isolated
Moraxella sp.	Aromatic	Aerobic	(0.1)	(29)	(7.2)	Unk.	Unk.	[351]	Settling tanks of sewage treatment plants, Ithaca and Marathon, NY, USA
Mycobacterium aromaticivorans	Aromatic	Aerobic	(0.8)	(28)	(6.8)	Unk.	Unk.	[390]	PAH contaminated soils of former oil-gasification site, Hilo, Big Island, HI, USA (19°49′20″ N, 155° 05′01″ W)
Mycobacterium crocinum	Aromatic	Aerobic	(0.8)	(28)	(6.8)	Unk.	Unk.	[390]	Pristine soil, central Oahu Island, HI, USA

(continued)

Table A.2 (continued)

Species	Hydrocarbon substrate	Oxygen utilization	Salinity condition (%)	Growth temperature (°C)	Growth pH	Biofilm forming	Surfactant producing	Reference	Isolation source/remarks
Mycobacterium eos	Aliphatic	Aerobic	(0.03)	(27)	(7.0–7.2)	Unk.	Unk.	[332]	In [391]
Mycobacterium frederiksbergense	Aromatic	Aerobic	(1.1)	15–37	Unk.	Unk.	Unk.	[392]	Coal tar-contaminated soil on the site of a former gas works at Frederiksberg, Denmark
Mycobacterium lacticola	Aliphatic	Aerobic	(0.03)	(27)	(7.0–7.2)	Unk.	Unk.	[332]	In [391]
Mycobacterium leprae	Mineral oil	Aerobic	(0.03)	(27)	(7.0–7.2)	Unk.	Unk.	[393]	Laboratory stock
Mycobacterium pallens	Aromatic	Aerobic	(0.8)	(28)	(6.8)	Unk.	Unk.	[390]	Pristine soil, central Oahu Island, HI, USA
Mycobacterium phlei	Mineral oil	Aerobic	(0.03)	(27)	(7.0–7.2)	Unk.	Unk.	[393]	Laboratory stock
Mycobacterium pyrenivorans	Aromatic	Aerobic	Unk.	24–37 (n/d)	(7.2)	Unk.	Unk.	[394]	PAH contaminated soil of a former coking plant, Übach-Palneberg, Germany
Mycobacterium rufum	Aromatic	Aerobic	(0.8)	(28)	(6.8)	Unk.	Unk.	[390]	PAH contaminated soils of former oil-gasification site, Hilo, Big Island, HI, USA (19° 49′ 20″ N, 155° 05′ 01″ W)
Mycobacterium rutilum	Aromatic	Aerobic	(0.8)	(28)	(6.8)	Unk.	Unk.	[390]	Pristine soil of urban park, Honolulu, Oahu Island, HI, USA
Mycobacterium smegmatis	Mineral oil	Aerobic	(0.03)	(27)	(7.0–7.2)	Unk.	Unk.	[393]	Laboratory stock
Mycobacterium sp.	Aromatic	Aerobic	(0.4)	(25)	(6.9)	Unk.	Unk.	[383]	Oil and PAH contaminated soil sites, Hamburg, Germany
Neptunomonas naphthovorans	Aromatic	Aerobic	1.8–7.0 (n/d)	4–24 (n/d)	6.5–8.5 (7.5)	Unk.	Unk.	[395]	Creosote contaminated sediment, Eagle Harbor Puget Sound, WA, USA
Nocardia asteroides	Aliphatic	Aerobic	(0.9)	(30)	Unk.	Unk.	Unk.	[382]	Hydrocarbon contaminated soils, north of Germany

(continued)

Table A.2 (continued)

Species	Hydrocarbon substrate	Oxygen utilization	Salinity condition (%)	Growth temperature (°C)	Growth pH	Biofilm forming	Surfactant producing	Reference	Isolation source/remarks
Nocardia corallina	Aliphatic/aromatic	Aerobic	Unk.	Unk.	Unk.	Unk.	Unk.	[396, 397]	Refinery soil Marcus Hook, PA, USA and pristine soil, Montgomery, AL, USA
Nocardia corynebacteroides	Aliphatic	Aerobic	(0.7)	(22–27)	(6.6)	Unk.	Pentasaccharide lipid	[398]	Oil contaminated soil
Nocardia sp.	Aromatic/aliphatic	Aerobic	(1)	(23)	Unk.	Unk.	Unk.	[327, 335]	Well water and core material contaminated with unleaded gasoline from a shallow costal aquifer, US Naval Weapons Station, Seal Beach, CA, USA; 220 strains isolated
Nocardioides aromaticivorans	Aromatic	Aerobic	Unk.	22–40 (30)	5–8 (7)	Unk.	Unk.	[399]	Contaminated surface water and sediments, Hikichi river, Kanagawa, Japan
Novosphingobium aromaticivorans	Aromatic	Aerobic	<3	(30–37)	Unk.	Unk.	Unk.	[400, 401]	Deep Cretaceous Atlantic coastal plain sediments; formerly *Sphingomonas aromaticivorans* [402]
Novosphingobium indicum	Aromatic	Aerobic	(0.2)	10–41 (25–30)	(7.5)	Unk.	Unk.	[403]	Deep seawater (4,546 m bleow the surface) on the Southwest Indian Ridge
Novosphingobium naphthalenivorans	Aromatic	Aerobic	unk.	15–37 (30)	5–8 (6.5–7.0)	Unk.	Unk.	[404]	Contaminated and pristine farmland soil and sediments Japan
Novosphingobium pentaromativorans	Aromatic	Aerobic	(0.3)	(25)	(7.2)	Unk.	Unk.	[405]	Estuarine sediment of Ulsan Bay, Republic of Korea

(continued)

Table A.2 (continued)

Species	Hydrocarbon substrate	Oxygen utilization	Salinity condition (%)	Growth temperature (°C)	Growth pH	Biofilm forming	Surfactant producing	Reference	Isolation source/remarks
Novosphingobium stygia	Aromatic	Aerobic	(0.5)	(30)	(7.1)	Unk.	Unk.	[400]	Deep Cretaceous Atlantic coastal plain sediments; formerly *Sphingomonas stygia* [402]
Novosphingobium subterranea	Aromatic	Aerobic	(0.5)	(30)	(7.1)	Unk.	Unk.	[400]	Deep Cretaceous Atlantic coastal plain sediments; formerly *Sphingomonas subterranea* [402]
Ochrobactrum anthropi	Aliphatic	Aerobic	(0.03)	(30)	(7.0–7.2)	Unk.	Unk.	[336]	Plant rhizosphere potted with crude oil contaminated soil of oil pumping sites, Zisterdorf, Lower Austria
Oleibacter marinus	Aliphatic	Aerobic	1–7 (n/d)	10–40 (25–30)	6–10 (6–7, 9)	Unk.	Unk.	[406]	Seawater, Pari Island near Jakarta, Indonesia
Oleiphilus messinensis	Aliphatic	Aerobic	0.06–20 (2.5–5.0)	10–37 (25–30)	Unk.	Yes	Unk.	[182]	Harbor seawater/sediment Messina, Sicily, Italy
Oleispira antarctica	Aliphatic	Aerobic	0.1–11.7	1–28	Unk.	Unk.	Unk.	[407]	Surface seawater, inlet Rod Bay, Ross Sea, Antarctica
Olivibacter oleidegradans	Diesel oil	Aerobic	(<2)	15–45 (30–37)	6.0–9.0 (6.5–7.0)	Unk.	Unk.	[408]	Hydrocarbon contaminated sediment filter of ex situ clean-up facility (former airbase), Hungary
Paenibacillus sp.	Aromatic	Aerobic	(0.4)	(25)	(6.9)	Unk.	Unk.	[326]	Tar oil contaminated soil. Hulin, Czech Republic
Pantoea agglomerans	Aliphatic	Aerobic	(0.5)	(30)	(7.0–7.2)	Unk.	Unk.	[336]	Plant rhizosphere potted with crude oil contaminated soil of oil pumping sites, Zisterdorf, Lower Austria

(continued)

Table A.2 (continued)

Species	Hydrocarbon substrate	Oxygen utilization	Salinity condition (%)	Growth temperature (°C)	Growth pH	Biofilm forming	Surfactant producing	Reference	Isolation source/remarks
Pasteurella sp.	Aromatic	Aerobic	(1.2)	(4–40)	Unk.	Unk.	Unk.	[375]	Activated sludge, Institute for Agrobiotechnology (IFA), Tulln Austria
Planococcus alkanoclasticus	Aliphatic	Aerobic	0.8–3.3 (n/d)	15–41 (n/d)	7.7–8.1 (n/d)	Unk.	Unk.	[409]	Fine sandy sediment from the intertidal zone of Stert Flats, Bridgewater Bay, Somerset, UK
Planomicrobium chinense	Aliphatic	Aerobic	(1–3)	30–50 (32)	6.4–7.8 (7.2)	Unk.	Yes	[410]	Diesel oil contaminated soil of petrol filling stations, Bilaspur, Chhattisgarh, India
Polaromonas naphthalenivorans	Aromatic	Aerobic	(0.1)	4–25 (20)	6.0–9.0 (7.0–7.5)	Unk.	Unk.	[411]	Surface freshwater sediment bathed in groundwater of naphthalene-rich coal tar waste
Polycyclovorans algicola	Aromatic	Aerobic	0.0–6.0 (3)	10–30 (30)	6.5–8.5 (8.3)	Unk.	Unk.	[412]	Nonaxenic laboratory culture of the marine diatom *Skeletonema costatum* CCAP1077/1C (origin, North Sea)
Porticoccus hydrocarbonoclasticus	Aromatic	Aerobic	(0.0–6.0)	10–37 (15)	6.5–9.0 (8.0)	Unk.	Unk.	[413]	Nonaxenic laboratory cultures of dinoflagellate *Ningulodinium polyedrum*
Proteus sp.	Diesel oil	Aerobic	(0.9)	(30)	Unk.	Unk.	Unk.	[333]	Rainforest soil, University of Port Harcourt, Nigeria

(continued)

Table A.2 (continued)

Species	Hydrocarbon substrate	Oxygen utilization	Salinity condition (%)	Growth temperature (°C)	Growth pH	Biofilm forming	Surfactant producing	Reference	Isolation source/remarks
Pseudoalteromonas sp.	Aromatic	Aerobic	(3.4)	4–37 (15–30)	(7.6)	Unk.	Unk.	[414]	Creosote contaminated sediment, Eagle Harbor Puget Sound, WA, USA
Pseudomonas aeruginosa	Aromatic/ aliphatic	Aerobic	(0.03)	(27)	6.0–9.5 (7.0–7.2)	Yes	Rhamnolipid	[335, 393, 415]	Well water and core material contaminated with unleaded gasoline from a shallow costal aquifer, US Naval Weapons Station, Seal Beach, CA, USA; 220 strains isolated
Pseudomonas alcaligenes	Aromatic/ aliphatic	Aerobic	(1)	(23)	Unk.	Unk.	Unk.	[335]	Well water and core material contaminated with unleaded gasoline from a shallow costal aquifer, US Naval Weapons Station, Seal Beach, CA, USA; 220 strains isolated
Pseudomonas anguiliseptica	Aliphatic	Aerobic	(0.03)	(30)	(7.0–7.2)	Unk.	Unk.	[336]	Plant rhizosphere potted with crude oil contaminated soil of oil pumping sites, Zisterdorf, Lower Austria
Pseudomonas boreopolis	Aliphatic	Aerobic	(0.03)	(30)	(7.0–7.2)	Unk.	Unk.	[336]	Plant rhizosphere potted with crude oil contaminated soil of oil pumping sites, Zisterdorf, Lower Austria

(continued)

Table A.2 (continued)

Species	Hydrocarbon substrate	Oxygen utilization	Salinity condition (%)	Growth temperature (°C)	Growth pH	Biofilm forming	Surfactant producing	Reference	Isolation source/remarks
Pseudomonas chlororaphis	Fuel oil	Aerobic	(0.4)	(24)	Unk.	Unk.	Unk.	[327]	Pristine temperate agricultural soil Montmirail, Marne, France
Pseudomonas cuatrocienegasensis	Aliphatic	Aerobic	(0.03)	(30)	(7.0–7.2)	Unk.	Unk.	[336]	Plant rhizosphere potted with crude oil contaminated soil of oil pumping sites, Zisterdorf, Lower Austria
Pseudomonas fluorescens	Fuel oil	Aerobic	(0.4)	(24)	Unk.	Unk.	Unk.	[327]	Pristine temperate agricultural soil Montmirail, Marne, France
Pseudomonas maltophila	Aromatic/ aliphatic	Aerobic	(1)	(23)	Unk.	Unk.	Unk.	[335]	Well water and core material contaminated with unleaded gasoline from a shallow costal aquifer, US Naval Weapons Station, Seal Beach, CA, USA; 220 strains isolated
Pseudomonas mendocina	Aromatic	Aerobic	(0.4)	(25)	(6.9)	Unk.	Unk.	[326]	Municipal wastewater plant, Gießen, Germany
Pseudomonas putida	Aromatic/ aliphatic	Aerobic/ anaerobic	0–2	10–35	Unk.	Yes	Biofilm, rhamnolipids	[335, 416–418]	Polluted creek Urbana, IL/ Bay Quinte, Kaje Intario, well water and core material contaminated with unleaded gasoline from a shallow costal aquifer, US Naval Weapons Station, Seal Beach, CA, USA; 220 strains isolated

(continued)

Table A.2 (continued)

Species	Hydrocarbon substrate	Oxygen utilization	Salinity condition (%)	Growth temperature (°C)	Growth pH	Biofilm forming	Surfactant producing	Reference	Isolation source/remarks
Pseudomonas pyocyaneus	Kerosene	Aerobic	(0.03)	(27)	6.0–9.5 (7.0–7.2)	Unk.	Unk.	[393]	Chicken, turkey, swine, water
Pseudomonas stutzeri	Aromatic/ aliphatic	Aerobic	(1)	(23)	Unk.	Unk.	Unk.	[335]	Well water and core material contaminated with unleaded gasoline from a shallow costal aquifer, US Naval Weapons Station, Seal Beach, CA, USA; 220 strains isolated
Pseudoxanthomonas sp.	Aliphatic	Aerobic	(0.03)	(30)	(7.0–7.2)	Unk.	Unk.	[336]	Plant rhizosphere potted with crude oil contaminated soil of oil pumping sites, Zisterdorf, Lower Austria
Ralstonia pickettii	Aromatic	Aerobic	(0.1)	(30)	(6.8)	Unk.	Unk.	[355]	Sandy aquifer near water table, General Technical Center, Warren, MI, USA
Rhodococcus aurantiacus	Aliphatic	Aerobic	(0.3)	(37)	(6.0)	Unk.	Yes	[419]	Unk.
Rhodococcus erythropolis	Aromatic/ aliphatic	Aerobic	(0.2–0.4)	(22–30)	(6.9–7.0)	Unk.	Trehalose lipids	[326, 398]	Municipal wastewater plant, Gießen, Germany/oil contaminated soil
Rhodococcus globerulus	Crude oil	Aerobic	Seawater	(30)	(7.8)	Unk.	Unk.	[420]	Unk.
Rhodococcus rhodochrous	Aromatic	Aerobic	(0.4)	(25)	(6.9)	Yes	Polysaccharides	[326, 420]	Industrial wastewater plant, Frankfurt, Germany
Rhodococcus sp.	Fuel oil	Aerobic	(0.4)	(24)	Unk.	Unk.	Unk.	[327]	Pristine temperate agricultural soil Montmirail, Marne, France

(continued)

Table A.2 (continued)

Species	Hydrocarbon substrate	Oxygen utilization	Salinity condition (%)	Growth temperature (°C)	Growth pH	Biofilm forming	Surfactant producing	Reference	Isolation source/remarks
Serratia sp.	Diesel oil	Aerobic	(0.9)	(30)	Unk.	Unk.	Unk.	[333]	Rainforest soil, University of Port Harcourt, Nigeria
Sphaerotilus natans	Petroleum	Aerobic	(1.3)	(15)	(7.0)	Unk.	Unk.	[421, 422]	Oil contaminted sediment from Colgate Creek, Baltimore Harbor, MD, USA
Sphingobium yanoikuyae	Aromatic	Aerobic	(1.1)	(30)	Unk.	Unk.	Unk.	[423]	Oil contaminated soil, Iowa City oil disposal facility; former *Sphingomonas yanoikuyae* [402]
Sphingomonas paucimobilis	Aromatic	Aerobic	(0.4)	(25)	(6.9)	Unk.	Unk.	[327, 383]	Oil and PAH contaminated soil sites, Hamburg, Germany/pristine temperate agricultural soil Montmirail, Marne, France
Sphingomonas sp.	Aromatic	Aerobic	(0.4)	(25)	(6.9)	Unk.	Unk.	[326]	Waste disposal site, leakage water, Reiskirchen, Germany
Sphingomonas spiritivorum	Fuel oil	Aerobic	(0.4)	(24)	Unk.	Unk.	Unk.	[327]	Pristine temperate agricultural soil Montmirail, Marne, France
Sphingopyxis macrogoltabida	Aliphatic	Aerobic	(0.03)	(30)	(7.0–7.2)	Unk.	Unk.	[336]	Root interior of plants potted with crude oil contaminated soil of oil pumping sites, Zisterdorf, Lower Austria

(continued)

Table A.2 (continued)

Species	Hydrocarbon substrate	Oxygen utilization	Salinity condition (%)	Growth temperature (°C)	Growth pH	Biofilm forming	Surfactant producing	Reference	Isolation source/remarks
Staphylococcus sp.	Aromatic	Aerobic	(0.6)	(28)	(7.0)	Unk.	Unk.	[424]	Creosote (PAH) contaminated soil
Stenotrophomonas maltophilia	Aromatic	Aerobic	(0.2)	(30)	Unk.	Unk.	Unk.	[425]	Soil from abandoned factory site near Port Melbourne, Australia; also *Xanthomonas maltophilia*
Streptomyces sp.	Aliphatic	Aerobic	(0.6)	(27)	(6.8)	Unk.	Unk.	[184, 426]	Oil polluted desert, Kuwait and oil contaminated soil, petrol bunk, Chennai, Tamil Nadu, India
Thalassolituus oleivorans	Aliphatic	Aerobic	0.5–5.7 (2.7)	4–30 (20–25)	7.5–9.0 (8.0)	Unk.	Unk.	[185]	Water/sediment from Harbour of Milazzo, Sicily, Italy
Thalassospira tepidiphila	Aromatic	Aerobic	2–14 (3)	8–45 (43)	5–10 (7)	Unk.	Unk.	[427]	Oil contaminated seawater
Thalassospira xianhensis	Aromatic	Aerobic	0.1–17.0 (3.0–6.0)	10–42 (30)	5.5–9.0 (7.2)	Unk.	Unk.	[428]	Oil-polluted saline soil in Xianhe, Shangdong Province, China
Thauera aromatica	Aromatic	Anaerobic	(0.1)	(30)	6.5–8.5	Yes	Unk.	[429, 430]	Gasoline contaminated soil, CA, USA
Thauera sp.	Aromatic	Aerobic/anaerobic	(0.3)	(30)	(7.0)	Unk.	Unk.	[431]	Anaerobic sludge from a wastewater treatment plant
Thermoleophilum album	Aliphatic	Aerobic	(1.1)	45–70 (58–62)	6.5–7.5 (7.0)	Unk.	Unk.	[432]	Thermal and nonthermal springs: Hot Springs, Ark, Faywood Hot Spring, NM, Yellowstone National Park, WY, Roanoke Rapids, NC all USA

(continued)

Table A.2 (continued)

Species	Hydrocarbon substrate	Oxygen utilization	Salinity condition (%)	Growth temperature (°C)	Growth pH	Biofilm forming	Surfactant producing	Reference	Isolation source/remarks
Thermoleophilum minutum	Aliphatic	Aerobic	(1.1)	45–70 (58–62)	6.0–7.0 (6.8)	Unk.	Unk.	[432, 433]	Yellowstone National Park, WY/Beaufort, NC, USA
Thermomicrobium fosteri	Aliphatic	Aerobic	(0.2)	42–70 (60)	(7.2–7.5)	Unk.	Unk.	[433, 434]	Mud sample from a littoral area near Beaufort, NC, USA
Thermomicrobium roseum	Aliphatic	Aerobic	(0.2)	70–85 (70)	(7.2–7.5)	Unk.	Unk.	[434]	Effluent hot spring Yellowstone National Park, WY/Beaufort, NC, USA
Thermus brockii	Aromatic/ aliphatic	Aerobic	(2.1)	70–80 (70)	(6–7)	Unk.	Unk.	[435]	Unk.
Thermus sp.	Aromatic	Aerobic	(2.1)	(60–70)	(6–7)	Unk.	Unk.	[435]	Unk.
Tranquillimonas alkanivorans	Aliphatic	Aerobic	1–13 (2)	10–50 (43)	6.5–9.5 (8.5)	Unk.	Unk.	[436]	Seawater Semarang Port, Indonesia
Tropicimonas isoalkanivorans	Aliphatic	Aerobic	1–6 (3–4)	10–46 (37)	5.5–7.6 (7.6)	Unk.	Unk.	[437]	Seawater Semarang Port, Indonesia
Tropicibacter naphthalenivorans	Aromatic	Aerobic	1–15 (5)	10–44 (37)	6.5–8.5 (7.6)	Unk.	Unk.	[438]	Seawater Semarang Port, Indonesia
Vibrio cyclotrophicus	Aromatic	Aerobic	1.75–10 (7)	4–37 (n/d)	6.5–9.5 (n/d)	Unk.	Unk.	[439]	Creosote contaminated sediment, Eagle Harbor Puget Sound, WA, USA
Vibrio spp.	Aromatic	Aerobic	(4.3)	4–28 (n.d.)	(7.6)	Unk.	Unk.	[440]	PAH contaminated marine sediment Eagle Harbor Puget Sound, WA, USA

Numbers in brackets are optimal growth conditions. Values without brackets indicate the growth range. When the publication does not distinguish between growth range and optimal growth, it was assumed that the given number represents optimal conditions and was placed in brackets
Unk unknown, *n/d* not determined

Table A.3 Values for the surface tension and the interfacial tension (γ) for certain microbial strains and pure surfactants

Origin	O_2	Surfactant	Incubation compound	ST w/o surfactant	ST reduction	γ w/o surfactant	γ reduction	γ compound	CMC	Ref.	Remark
Unknown	Unknown	Glycolipids	n/d	n/d	n/d	n/a	3	n/a	n/a	[441]	
Acinetobacter calcoaceticus strains	Aerobic	Capsular polysaccharide	Unknown	n/d	n/d	n/a	n/d	n/a	n/a	[442]	
Acinetobacter clacoaceticus RAG-1	Aerobic	EF-RAG	*n*-Hexadecane	n/d	n/d	n/a	n/d	n/a	n/a	[172]	
Acinetobacter radioresistens	Aerobic	Alasan	Ethanol	69.1	41.6	n/a	n/d	n/a	0.2 mg l^{-1}	[443, 444]	
Acinetobacter sp.	Aerobic	Particulate	*n*-Hexadecane	n/d	n/d	n/a	n/d	n/a	n/a	[445]	
***Alcaligenes faecalis* MS-103**	**Aerobic**	**n/d**	**Molasses**	**63.0**	**20**	**24**	**<1**	**Not given**	**n/d**	[446]	
Alcanivorax borkumensis MM1	Aerobic	Glycolipid	Pyruvate and *n*-alkane mix	72	30	n/d	<5	*n*-Hexadecane	n/d	[337, 447]	
Aeribacillus pallidus YM-1	Aerobic	Heteropolysaccharide	Crude oil	70.5	45.7	n/d	n/d	n/a	n/d	[448]	
Arthrobacter sp. DSM2567	Aerobic	Glycolipid	Glucose + yeast extract	72.0	33	41	1	*n*-Hexadecane	1–50 mg l^{-1}	[449]	
Arthrobacter sp. RAG-1	Aerobic	EF-RAG	*n*-hexadecane	n/d	n/d	n/d	6.9	Crude oil	1–10 g l^{-1}	[450]	
Bacillus cereus IAF 346	Aerobic	Monoglyceride	Sucrose	n/d	28	n/a	n/a	n/a	n/a	[451]	
Bacillus circulans	Aerobic	n/d	*n*-Hexadecane	68.0	37.2	n/a	n/d	n/a	n/d	[452]	
Bacillus licheniformis	Aerobic	n/d	*n*-Hexadecane	68.0	37.4	n/a	n/d	n/a	n/d	[452]	
***Bacillus licheniformis* 86**	**Aerobic**	**Lipopeptides**	Glucose	n/d	**27**	n/a	**0.36**	***n*-Hexadecane**	**n/d**	[453]	
Bacillus licheniformis BAS50	Aerobic	Lichenysin A	Glucose	72.0	28.3	n/a	n/d	n/a	12 mg l^{-1}	[76]	
Bacillus licheniformis BAS50	Anaerobic	Lichenysin A	Glucose	72.0	35	n/a	n/d	n/a	12 mg l^{-1}	[76]	
Bacillus mojavensis JF2	Aerobic	Lichenysin	Carbohydrates	77	26.5	9.5	1	*n*-Hexadecane	n/d	[454, 455]	Renamed *B. licheniformis*
***Bacillus mojavensis* JF2**	**Aerobic**	**Surfactin**	**Glucose**	**n/d**	**n/d**	**32**	**0.006**	***n*-Decane**	**10 mg l^{-1}**	[212, 455]	**Strong pH dependence**
Bacillus mojavensis JF2	Anaerobic	n/d	Carbohydrates	70–74	27–31	n/a	n/d	n/a	4 mg l^{-1}	[202]	Nitrate required
Bacillus mojavensis JF2 mutant	Aerobic	n/d	Glucose	72.1	26.5	n/a	n/d	n/a	10 mg l^{-1}	[213, 455]	
Bacillus sp ?	Aerobic	n/d	n/a	n/d	27	n/d	1	n/d	5 mg l^{-1}	[456]	
Bacillus stearothermophilus	Aerobic	n/d	Crude oil	n/d	45	n/d	n/d	n/a	n/a	[457]	
Bacillus subtilis	Aerobic	n/d	*n*-Hexadecane	68.0	39.6	n/a	n/d	n/a	n/d	[452]	

(continued)

Table A.3 (continued)

Origin	O_2	Surfactant	Incubation compound	ST w/o surfactant	ST reduction	γ w/o surfactant	γ reduction	γ compound	CMC	Ref.	Remark
Bacillus subtilis	Aerobic	Purified surfactin	n/a	72.0	29	n/a	n/d	n/a	50–500 mg l^{-1}	[458]	
Bacillus subtilis	**Aerobic**	**Surfactin**	***n*-Hexadecane**	**n/d**	**27**	**n/d**	**<1**	***n*-Hexadecane**	**25 mg l^{-1}**	[201]	**pH dependent**
Bacillus subtilis C9	Aerobic	C9-BS	Glucose	72	28	n/a	n/d	n/a	n/d	[205]	
***Bacillus subtilis* LB5a**	**Aerobic**	**n/d**	**Cassava waste**	**n/d**	**26**	**n/d**	**0.97**	***n*-Hexadecane**	**33 mg l^{-1}**	[459]	
Bio-synthetic	**n/a**	**Di-rhamnolipid methyl ester**	**n/a**	**n/d**	**31**	**n/d**	**<0.1**	***n*-Hexadecane**	**0.04 mg l^{-1}**	[173]	
Bio-synthetic	n/a	Di-rhamnolipid acid	n/a	n/d	36	n/d	5	*n*-Hexadecane	0.1 mg l^{-1}	[173]	
Bio-synthetic	n/a	Trehalose-6,6′-dicorynomycolates	n/a	n/a	n/d	21	16	Paraffin S	n/d	[460, 461]	
Bio-synthetic	n/a	Cellubiose lipids	n/a	n/a	n/d	21	3	Paraffin S	n/d	[460, 461]	
Bio-synthetic	n/a	Sophorose lipids	n/a	n/a	n/d	21	2	Paraffin S	n/d	[460, 461]	
Bio-synthetic	n/a	Rhamnolipids	n/a	n/a	n/d	21	4	Paraffin S	n/d	[460, 461]	
Bio-synthetic	n/a	Trehalose-2,3,4,2'-tetraesters	n/a	n/a	n/d	21	2	Paraffin S	n/d	[460, 461]	
Bio-synthetic	**n/a**	**1:1 mono-:dirhamnolipid**	**n/a**	**n/a**	**n/d**	**n/d**	**0.025**	**Toluene**	**10 mg l^{-1}**	[208]	**Optimal salinity 3 %**
Candida bombicola	Aerobic	Sophorolipids	Sunflower oil	n/d	31	n/a	n/d	n/a	n/d	[462]	
Candida lipolytica	Aerobic	Liposan	n/a	n/d	n/d	n/a	n/d	n/a	n/d	[463]	
Clostridium pasteurianum	Anaerobic	Unknown	n/a	72	55	n/a	n/d	n/a	n/d	[456]	
Corynebacterium hydrocarboclastus	Aerobic	Polymer	*n*-Heptadecane	n/d	40	n/a	n/a	n/a	2.4 g l^{-1}	[357]	
Corynebacterium lepus	Aerobic	Lipids	Kerosene	72	30	n/a	n/d	n/a	n/d	[464]	
Enterobacter cloacae	Aerobic		Olive oil	71	53.6	18	15.2	Crude oil	n/d	[177]	
Enterobacter/ Pseudomonas	**Aerobic**		**Olive oil**	**71**	**31.7**	**18**	**0.65**	**Crude oil**	**n/d**	[177]	
Geobacillus pallidus	Aerobic	Unknown	n/a	70.5	42.5	n/a	n/d	n/a	n/d	[298]	
Haloarcula sp. D21	Aerobic	n/d	Diesel oil	n/d	26.2	n/a	n/d	n/a	n/d	[465]	
Halomonas eurihalina	Aerobic	EPS	Peptone and several oils	n/d	n/d	n/a	n/d	n/a	n/a	[466]	
Halovivax sp. A21	Aerobic	n/d	Diesel oil	n/d	28.4	n/a	n/d	n/a	n/d	[465]	
Marinobacter hydrocarbonoclasticus	Aerobic	EPS	*n*-Hexadecane	n/a	n/d	42	18	*n*-Hexadecane	n/a	[157]	
Methanobacterium thermoautotrophicum	Anaerobic	No surfactant	Yeast extract	65	39.5	23	20	*n*-Hexadecane	n/a	[467]	
Microbacterium G35-I	Aerobic	n/d	*n*-Hexadecane	68.0	37.1	n/a	n/d	n/a	n/d	[452]	

(continued)

Table A.3 (continued)

Origin	O$_2$	Surfactant	Incubation compound	ST w/o surfactant	ST reduction	γ w/o surfactant	γ reduction	γ compound	CMC	Ref.	Remark
Nocardia corynebacteroides	Aerobic	Pentasaccharide lipid	*n*-Alkane	72	26	43	<1	*n*-Hexadecane	30 mg l^{-1}	[398]	
Pseudomonas aeroginosa	Aerobic	n/d	petroleum, octane, toluene	68.0	35.1	n/a	n/d	n/a	100 mg l^{-1}	[267]	Stable under reservoir conditions
Pseudomonas aeroginosa	Aerobic	n/d	*n*-Hexadecane	68.0	33.5–39.6	n/a	n/d	n/a	29.5–68.9 g l^{-1}	[452]	
Pseudomonas aeruginosa	Aerobic	Rhamnolipid	Peptone & glucose	66	29	n/a	n/d	n/a	40 mg l^{-1}	[468]	pH effect
***Pseudomonas aeruginosa* 44T1**	**Aerobic**	**Rhamnolipid**	**Glucose**	**n/d**	**25**	**n/d**	**0.2-1**	**Kerosene**	**11 ppm**	[469]	
Pseudomonas aeruginosa mutant	Aerobic	Rhamnolipids	Yeast extract	n/d	28	n/a	n/d	n/a	9 **mg l^{-1}**	[470]	
***Pseudomonas aeruginosa* mutant**	**Aerobic**	**Rhamnolipids**	**Waste oils**	**41**	**29**	**21**	**<1**	***n*-Hexadecane**	**57 mg l^{-1}**	[471]	
Pseudomonas BOP 100	Aerobic	Rhamnolipid RB-Me	*n*-Hexadecane	n/d	30	n/d	0.1	*n*-Octane	0.4 g l^{-1}	[472]	
Pseudomonas fluorescens	Aerobic	n/d	*n*-Hexadecane	68.0	35.1	n/a	n/d	n/a	n/d	[452]	
Pseudomonas putida 21BN	Aerobic	Rhamnolipids	*n*-Hexadecane	71	29	n/a	n/d	n/a	n/d	[473]	
Pseudomonas sp.	Aerobic	Rhamnolipids	Olive oil mill effluent	42	30	n/d	n/d	n/a	n/a	[474]	
Pseudomonas sp.	Aerobic		Olive oil	71	39.4	18	1.8	Crude oil	n/d	[177]	
***Pseudomonas* sp. DSM2874**	**Aerobic**	**Rhamnolipids**	**n/a**	**n/d**	**26–30**	**n/d**	**<1.4**	***n*-Hexadecane**	**10-200 mg l^{-1}**	[475]	
Renibacterium salmoninarum 27BN	Aerobic	Rhamnolipid	*n*-Hexadecane	71.0	21–25	n/a	n/d	n/a	n/d	[350]	
***Rhodococcus aurantiacus* 80001**	**Aerobic**	**n/d**	***n*-Alkane**	**n/d**	**26**	**n/d**	**0.35**	***n*-Hexadecane**	**n/d**	[419]	
Rhodococcus erythropolis ATCC4277	Aerobic	n/d	Yeast extract and glycerol	n/a	n/d	43	15	*n*-Hexadecane	n/d	[476]	
***Rhodococcus erythropolis* DSM43215**	**Aerobic**	**Trehalose-2,20,3,4-tetraester**	***n*-Alkanes**	**72**	**26**	**43**	**<1**	***n*-Hexadecane**	**15 mg l^{-1}**	[398]	
Rhodococcus erythropolis DSM43215	Aerobic	Trehalose lipids	*n*-Alkanes	72.0	30	44	18	*n*-Alkanes	0.7–165 mg l^{-1}	[477]	
***Rhodococcus erythropolis* DSM43215**	**Aerobic**	**Phosphatidylethanolamines**	***n*-Alkanes**	**72.0**	**32**	**44**	**<1**	***n*-Alkanes**	**5–30 mg l^{-1}**	[477]	
***Rhodococcus ruber* Z25**	**Aerobic/ anaerobic**	**n/d**	**Paraffin**	**n/a**	**n/a**	**n/d**	**1**	***n*-Hexadecane**	**75 mg l^{-1}**	[317]	
Rhodococcus sp. 51T7	Aerobic	n/d	*n*-Hexadecane	n/a	30	n/a	n/d	n/a	n/a	[478]	

(continued)

Table A.3 (continued)

Origin	O_2	Surfactant	Incubation compound	ST w/o surfactant	ST reduction	γ w/o surfactant	γ reduction	γ compound	CMC	Ref.	Remark
***Rhodococcus* sp. H13-A**	**Aerobic**	**Octaacyl-trehaloses**	**n/a**	**n/a**	**n/d**	**n/a**	**0.02–0.00005**	***n*-Decane**	**1.0–1.5 g l^{-1}**	[156]	**0.00006 with *n*-pentanol**
Rhodococcus terrae 70006	Aerobic	n/d	n/a	n/a	30	n/a	3	*n*-Hexadecane	n/d	[479]	
synthetic	n/a	Sodium dodecyl sulfate	n/a	73	32	49	10	*n*-Hexadecane	8 mM	[480]	
synthetic	**n/a**	**Rhamnolipid-synthetic mix**	**n/a**	**n/a**	**n/d**	**n/d**	**0.029**	***n*-Decane**	**<1 mg l^{-1}**	[208]	
Torulopsis apicola	**Aerobic**	**Sophorolipids**	***n*-Hexadecane**	**72.0**	**22.7**	**n/d**	**<0.9**	***n*-Decane**	**n/d**	[481, 482]	**Yeast**
Torulopsis apicola	Aerobic	Sophorolipids	*n*-Hexadecane	70	26	n/d	9	*n*-Decane	60 mg l^{-1}	[483]	Yeast
Torulopsis bombicola	Aerobic	Lactonic sophorolipid	Food grade oil	>51.0	31	n/d	1.5	*n*-Decane	82 mg l^{-1}	[484, 485]	\$2.75 kg^{-1} production cost
Torulopsis candida	Aerobic	Glycolipids	*n*-Hexadecane	72.0	50.1	n/d	n/d	*n*-Decane	n/d	[481]	Yeast
Torulopsis ernobii	Aerobic	Glycolipids	*n*-Hexadecane	72.0	72	n/d	52.0	*n*-Decane	n/d	[481]	Yeast
Torulopsis fumata	Aerobic	Glycolipids	*n*-Hexadecane	72.0	50.1	n/d	n/d	*n*-Decane	n/d	[481]	Yeast
Torulopsis glabrata	Aerobic	Glycolipids	*n*-Hexadecane	72.0	39.1	n/d	32.0	*n*-Decane	n/d	[481]	Yeast
Torulopsis petrophilum	Aerobic	Lactonic sophorolipid	Sunflower oil	55.0	43	n/a	n/d	n/a	n/d	[484, 485]	Yeast

Bold indicates γ values below 1, which is necessary for effective oil displacement. O_2 indicates oxygen utilization
CMC critical micelle concentration; *Ref.* reference, *w/o* without, *n/d* not determined, *n/a* not applicable

References

1. U.S. Energy Information Administration, U.S. Department of Energy (2010) International Energy Outlook. Office of Integrated Analysis and Forecasting: Washington, DC
2. BP p.l.c. Statistical Review of World Energy 2013. 22 June 2013. Available from: http://www.bp.com/content/dam/bp/excel/Statistical-Review/statistical_review_of_world_energy_2013_workbook.xlsx.
3. U.S. Department of Energy (2006) Undeveloped domestic oil resources. U.S. Department of Energy: Washington, DC, USA
4. Zobell CE (1946) Bacteriological process for treatment of fluid-bearing earth formations. United States, US Patent 2413278
5. Brown LR (2010) Microbial enhanced oil recovery (MEOR). Curr Opin Microbiol 13:316–320
6. Lazar I (1991) MEOR field trials carried out over the world during the last 35 Years. In: Erle CD (ed) Developments in petroleum science. Elsevier, Amsterdam
7. Youssef N, Elshahed MS, McInerney MJ (2009) Microbial processes in oil fields: culprits, problems, and opportunities. In: Allen I, Laskin SS, Geoffrey MG (eds) Advances in applied microbiology. Academic Press, New York
8. U.S. Energy Information Administration (2013) Spot Prices for Crude Oil and Petroleum Products. http://www.eia.gov/dnav/pet/pet_pri_spt_s1_m.htm. Accessed 9–13 Jan 2013
9. Volk H, Hendry P (2010) 3° oil recovery: fundamental approaches and principles of microbially enhanced oil recovery. In: Timmis KN (ed) Handbook of hydrocarbon and lipid microbiology. Springer, Berlin
10. Moses V, Springham DG (1982) Bacteria and the enhancement of oil recovery. Applied Science Publishers, London
11. Meyer JP (2010) Summary of carbon dioxide enhanced oil recovery (CO_2EOR) injection well technology. American Petroleum Institute, Plano, TX, USA
12. Bastin ES (1926) The problem of the natural reduction of sulphates. Bull Am Assoc Petrol Geol 10:1270–1299
13. Galushko AS, Rozanova EP (1991) *Desulfobacterium cetonicum* spec. nov., a sulfate-reducing bacterium oxidizing fatty acids and ketones. Microbiology (Moscow, Russ. Fed.), 60:102–107
14. Rozanova EP, Nazina TN, Galushko AS (1988) Isolation of a new genus of sulfate-reducing bacteria and description of a new species of this genus, *Desulfomicrobium apsheronum* gen. nov., sp. nov. Microbiology (Moscow, Russ. Fed.) 57: 634–641
15. Ng TK, Weimer PJ, Gawel LJ (1989) Possible nonanthropogenic origin of two methanogenic isolates from oil-producing wells in the san miguelito field, ventura county, California. Geomicrobiol J 7:185–192
16. Voordouw G, Armstrong SM, Reimer MF, Fouts B et al (1996) Characterization of 16S rRNA genes from oil field microbial communities indicates the presence of a variety of sulfate-reducing, fermentative, and sulfide-oxidizing bacteria. Appl Environ Microbiol 62:1623–1629
17. L'Haridon S, Reysenbacht AL, Glenat P, Prieur D, Jeanthon C (1995) Hot subterranean biosphere in a continental oil reservoir. Nature 377:223–224
18. Magot M, Ollivier B, Patel BKC (2000) Microbiology of petroleum reservoirs. Antonie Leeuwenhoek 77:103–116
19. Orphan VJ, Taylor LT, Hafenbradl D, Delong EF (2000) Culture-dependent and culture-independent characterization of microbial assemblages associated with high-temperature petroleum reservoirs. Appl Environ Microbiol 66:700–711
20. Birkeland N-K (2004) Chapter 14: The microbial diversity of deep subsurface oil reservoirs. In: Vazquez-Duhalt R, Quintero-Ramirez R (eds) Studies in Surface Science and Catalalysis. Elsevier, Amsterdam

21. Grassia GS, McLean KM, Glénat P, Bauld J, Sheehy AJ (1996) A systematic survey for thermophilic fermentative bacteria and archaea in high temperature petroleum reservoirs. FEMS Microbiol Ecol 21:47–58
22. Magot M (1996) Similar bacteria in remote oil fields. Nature 379:681
23. Head IM, Jones DM, Larter SR (2003) Biological activity in the deep subsurface and the origin of heavy oil. Nature 426:344–352
24. Sunde E, Torsvik T (2005) Microbial control of hydrogen sulfide production in oil reservoirs. In: Ollivier B, Magot M (eds) Petroleum microbiology. ASM Press, Washington, DC
25. Venzlaff H, Enning D, Srinivasan J, Mayrhofer KJJ et al (2013) Accelerated cathodic reaction in microbial corrosion of iron due to direct electron uptake by sulfate-reducing bacteria. Corros Sci 66:88–96
26. Matthews JD (2008) The origin of oil—a creationist answer. Ans Res J 1:145–168
27. Tissot BP, Welte HD (1984) Petroleum formation and occurrence, 2nd edn. Springer, Berlin
28. Magot M (2005) Indigenous microbial communities in oil fields. In: Ollivier B, Magot M (eds) Petroleum microbiology. ASM Press, Washington, DC
29. Stetter KO, Huber R, Blochl E, Kurr M et al (1993) Hyperthermophilic archaea are thriving in deep North Sea and Alaskan oil reservoirs. Nature 365:743–745
30. Jones DM, Head IM, Gray ND, Adams JJ et al (2008) Crude-oil biodegradation via methanogenesis in subsurface petroleum reservoirs. Nature 451:176–180
31. Sunde E (2006) Method of treating a hydrocarbon bearing formation. United States, US Patent 7124817
32. Hyne NJ (2001) Nontechnical guide to petroleum geology, exploration, drilling, and production, 2nd edn. PennWell Corporation Oklahoma, Oklahoma
33. Nourani M, Panahi H, Mohebbi A, Haghighi M, Roostaazad R, Biria D (2007) Laboratory studies of MEOR in the micromodel as a fractured system. In: Eastern Regional Meeting, SPE Publications, Lexington
34. Abtahi N, Roostaazad R, Ghadiri F (2003) Biosurfactant production. In: MEOR for improvement of Iran's Oil reservoirs' production experimental approach. In: SPE international improved oil recovery conference in Asia Pacific. SPE Publications, Kuala Lumpur
35. Sayyouh MH (2002) Microbial enhanced oil recovery: research studies in the Arabic area during the last ten years. In: SPE/DOE improved oil recovery symposium. SPE Publications, Tulsa
36. Kadarwati S, Udiharto M, Hadi N, Doria I (1999) Selected Indonesian microbes potentials for MEOR. In: SPE Asia Pacific improved oil recovery conference. SPE Publications, Kuala Lumpur
37. Yusuf A, Sri K, Nurkamelia, Sumaryana (1999) Field test of the indigenous microbes for oil recovery, Ledok Field, Central Java. In: SPE Asia Pacific improved oil recovery conference. SPE Publications, Kuala Lumpur, Malaysia. SPE accession number 57309, doi: 10.2118/57309-MS
38. Sugihardjo, Legowo EH, Pratomo, Pratomo SW (1999) Microbial core flooding experiments using indigenous microbes. In: SPE Asia Pacific improved oil recovery conference. SPE Publications, Kuala Lumpur, Malaysia. SPE accession number 57306, doi: 10.2118/57306-MS
39. Sugai Y, Kishita A, Hong CX, Enomoto H, Chida T, Zhou SC (1999) Laboratory investigation of polymer producer TU-15A for MEOR field tests. In: SPE Asia Pacific oil and gas conference and exhibition. SPE Publications, Jakarta, pp. 1–13
40. AlMaghrabi I, Chaalal O, Islam MR (1998) Thermophilic bacteria in UAE environment can enhance biodegradation and mitigate wellbore problems. In: Abu Dhabi international petroleum exhibition and conference. SPE Publications, Abu Dhabi
41. Ohno K, Maezumi S, Sarma HK, Enomoto H et al (1999) Implementation and performance of a microbial enhanced oil recovery field pilot in Fuyu oilfield, China. In: SPE Asia Pacific oil and gas conference and exhibition. SPE Publications, Jakarta
42. Lewin A, Johansen J, Wentzel A, Kotlar HK, Drabløs F, Valla S (in press) The microbial communities in two apparently physically separated deep subsurface oil reservoirs show extensive DNA sequence similarities. Environ Microbiol. doi:10.1111/1462-2920.12181 (in press)

43. Aitken CM, Jones DM, Larter SR (2004) Anaerobic hydrocarbon biodegradation in deep subsurface oil reservoirs. Nature 431:291–294
44. Wilhelms A, Larter SR, Head I, Farrimond P, di-Primio R, Zwach C (2001) Biodegradation of oil in uplifted basins prevented by deep-burial sterilization. Nature 411:1034–1037
45. da Cunha C, Rosado A, Sebastián G, Seldin L, von der Weid I (2006) Oil biodegradation by *Bacillus* strains isolated from the rock of an oil reservoir located in a deep-water production basin in Brazil. Appl Microbiol Biotechnol 73:949–959
46. Spark I, Patey I, Duncan B, Hamilton A, Devine C, McGovern-Traa C (2000) The effects of indigenous and introduced microbes on deeply buried hydrocarbon reservoirs, North Sea. Clay Miner 35:5–12
47. Beeder J, Nilsen RK, Thorstenson T, Torsvik T (1996) Penetration of sulfate reducers through a porous North Sea oil reservoir. Appl Environ Microbiol 62:3551–3553
48. Nilsen RK, Beeder J, Thorstenson T, Torsvik T (1996) Distribution of thermophilic marine sulfate reducers in North Sea oil field waters and oil reservoirs. Appl Environ Microbiol 62:1793–1798
49. Slobodkin AI, Jeanthon C, L'Haridon S, Nazina T, Miroshnichenko M, Bonch-Osmolovskaya E (1999) Dissimilatory reduction of Fe(III) by thermophilic bacteria and archaea in deep subsurface petroleum reservoirs of Western Siberia. Curr Microbiol 39:99–102
50. Kotlar HK, Lewin A, Johansen J, Throne-Holst M et al (2011) High coverage sequencing of DNA from microorganisms living in an oil reservoir 2.5 kilometres subsurface. Environ Microbiol Rep 3:674–681
51. Basso O, Lascourrèges J-F, Jarry M, Magot M (2005) The effect of cleaning and disinfecting the sampling well on the microbial communities of deep subsurface water samples. Environ Microbiol 7:13–21
52. Park CB, Clark DS (2002) Rupture of the cell envelope by decompression of the deep-sea methanogen *Methanococcus jannaschii*. Appl Environ Microbiol 68:1458–1463
53. Belyaev SS, Wolkin R, Kenealy WR, DeNiro MJ, Epstein S, Zeikus JG (1983) Methanogenic bacteria from the bondyuzhskoe oil field: general characterization and analysis of stable-carbon isotopic fractionation. Appl Environ Microbiol 45:691–697
54. Schippers A, Neretin LN, Kallmeyer J, Ferdelman TG et al (2005) Prokaryotic cells of the deep sub-seafloor biosphere identified as living bacteria. Nature 433:861–864
55. Youssef N, Simpson DR, Duncan KE, McInerney MJ et al (2007) In situ biosurfactant production by *Bacillus* strains injected into a limestone petroleum reservoir. Appl Environ Microbiol 73:1239–1247
56. Brioukhanov AL, Netrusov AI (2004) Catalase and superoxide dismutase: distribution, properties, and physiological role in cells of strict anaerobes. Biochemistry (Mosc.) 69:949–962
57. Cypionka H (2000) Oxygen respiration by *Desulfovibrio* species. Annu Rev Microbiol 54:827–848
58. Voordouw G (1995) The genus *Desulfovibrio*: the centennial. Appl Environ Microbiol 61:2813–2819
59. Colin Y, Goñi-Urriza M, Caumette P, Guyoneaud R (2013) Combination of high throughput cultivation and dsrA sequencing for assessment of sulfate-reducing bacteria diversity in sediments. FEMS Microbiol Ecol 83:26–37
60. Thamdrup B, Rossello-Mora R, Amann R (2000) Microbial manganese and sulfate reduction in black sea shelf sediments. Appl Environ Microbiol 66:2888–2897
61. Sowers KR, Boone JE, Gunsalus RP (1993) Disaggregation of *Methanosarcina* spp. and growth as single cells at elevated osmolarity. Appl Environ Microbiol 59:3832–3839
62. Galushko AS, Rozanova EP, Ivanova AE (1993) Sulfate reduction in waters injected into oil strata. Microbiology (Moscow, Russ. Fed.) 62:653–656
63. Hungate RE (1969) A roll tube method for cultivation of strict anaerobes. In: Norris R, Ribbons DW (eds) Methods in microbiology. Academic Press, New York

64. Ali SA, Hasan F, Shah Z, Kanwal N, Zeb S (2013) Biodegradation of natural and synthetic rubbers: a review. Int Biodeterior Biodegrad 83:145–157
65. Widdel F (2010) Cultivation of anaerobic microorganisms with hydrocarbons as growth substrates. In: Timmis KN (ed) Handbook of hydrocarbon and lipid microbiology. Springer, Berlin
66. Balch WE, Fox GE, Magrum LJ, Woese CR, Wolfe RS (1979) Methanogens: reevaluation of a unique biological group. Microbiol Rev 43:260–296
67. Widdel F, Bak F (1992) Gram-negative mesophilic sulfate-reducing bacteria. In: Dworkin M (ed) The prokaryotes. Springer, New York
68. Mahlert F, Bauer C, Jaun B, Thauer R, Duin E (2002) The nickel enzyme methyl-coenzyme M reductase from methanogenic archaea: In vitro induction of the nickel-based MCR-ox EPR signals from MCR-red2. J Biol Inorg Chem 7:500–513
69. Hungate RE (1950) The anaerobic mesophilic cellulolytic bacteria. Bacteriol Rev 14:1–49
70. Wolfe RS, Metcalf WW (2010) A vacuum-vortex technique for preparation of anoxic solutions or liquid culture media in small volumes for cultivating methanogens or other strict anaerobes. Anaerobe 16:216–219
71. Jones GA, Pickard MD (1980) Effect of titanium (III) citrate as reducing agent on growth of rumen bacteria. Appl Environ Microbiol 39:1144–1147
72. Leadbetter JR, Breznak JA (1996) Physiological ecology of *Methanobrevibacter cuticularis* sp. nov. and *Methanobrevibacter curvatus* sp. nov., isolated from the hindgut of the termite *Reticulitermes flavipes*. Appl Environ Microbiol 62:3620–3631
73. Jain DK, Collins-Thompson DL, Lee H, Trevors JT (1991) A drop-collapsing test for screening surfactant-producing microorganisms. J Microbiol Methods 13:271–279
74. Amann RI, Ludwig W, Schleifer KH (1995) Phylogenetic identification and in situ detection of individual microbial cells without cultivation. Microbiol Rev 59:143–169
75. McInerney MJ, Knapp RM, Duncan K, Simpson DR et al (2007) Development of an in situ biosurfactant production technology for enhanced oil recovery. University of Oklahoma.
76. Yakimov MM, Kröger A, Slepak TN, Giuliano L, Timmis KN, Golyshin PN (1998) A putative lichenysin A synthetase operon in *Bacillus licheniformis* initial characterization. Biochimica et Biophysica Acta (BBA)—Gene Structure and Expression, 1399:141–153
77. Philippot L, Piutti S, Martin-Laurent F, Hallet S, Germon JC (2002) Molecular analysis of the nitrate-reducing community from unplanted and maize-planted soils. Appl Environ Microbiol 68:6121–6128
78. Palmer K, Biasi C, Horn MA (2012) Contrasting denitrifier communities relate to contrasting N_2O emission patterns from acidic peat soils in arctic tundra. ISME J 6:1058–1077
79. Throbäck IN, Enwall K, Jarvis Å, Hallin S (2004) Reassessing PCR primers targeting *nirS*, *nirK* and *nosZ* genes for community surveys of denitrifying bacteria with DGGE. FEMS Microbiol Ecol 49:401–417
80. Rich JJ, Heichen RS, Bottomley PJ, Cromack K, Myrold DD (2003) Community composition and functioning of denitrifying bacteria from adjacent meadow and forest soils. Appl Environ Microbiol 69:5974–5982
81. Wagner M, Loy A, Klein M, Lee N et al (2005) Functional marker genes for identification of sulfate-reducing prokaryotes. In: Leadbetter JR (ed) Methods in enzymology. Academic Press. New York
82. Pester M, Bittner N, Deevong P, Wagner M, Loy A (2010) A 'rare biosphere' microorganism contributes to sulfate reduction in a peatland. ISME J 4:1591–1602
83. Hales BA, Edwards C, Ritchie DA, Hall G, Pickup RW, Saunders JR (1996) Isolation and identification of methanogen-specific DNA from blanket bog peat by PCR amplification and sequence analysis. Appl Environ Microbiol 62:668–675
84. Luton PE, Wayne JM, Sharp RJ, Riley PW (2002) The *mcrA* gene as an alternative to 16S rRNA in the phylogenetic analysis of methanogen populations in landfill. Microbiology 148:3521–3530

85. Steinberg LM, Regan JM (2009) *mcrA*-Targeted real-time quantitative PCR method to examine methanogen communities. Appl Environ Microbiol 75:4435–4442
86. von Netzer F, Pilloni G, Kleindienst S, Krüger M et al (2013) Enhanced gene detection assays for fumarate-adding enzymes allow uncovering anaerobic hydrocarbon degraders in terrestrial and marine systems. Appl Environ Microbiol 79:543–552
87. Kuntze K, Kiefer P, Baumann S, Seifert J et al (2011) Enzymes involved in the anaerobic degradation of meta-substituted halobenzoates. Mol Microbiol 82:758–769
88. Kuntze K, Shinoda Y, Moutakki H, McInerney MJ et al (2008) 6-Oxocyclohex-1-ene-1-carbonyl-coenzyme A hydrolases from obligately anaerobic bacteria: characterization and identification of its gene as a functional marker for aromatic compounds degrading anaerobes. Environ Microbiol 10:1547–1556
89. Kepner RL, Pratt JR (1994) Use of fluorochromes for direct enumeration of total bacteria in environmental samples: past and present. Microbiol Rev 58:603–615
90. Noble RT, Fuhrman JA (1998) Use of SYBR Green I for rapid epifluorescence counts of marine viruses and bacteria. Aquat Microb Ecol 14:113–118
91. Pernthaler J, Glöckner F-O, Schönhuber W, Amann R (2001) Fluorescence in situ hybridization (FISH) with rRNA-targeted oligonucleotide probes. In: Paul JH (ed) Methods in microbiology. Academic Press, New York
92. Pernthaler A, Pernthaler J, Amann R (2002) Fluorescence in situ hybridization and catalyzed reporter deposition for the identification of marine bacteria. Appl Environ Microbiol 68:3094–3101
93. Zhang T, Fang HP (2006) Applications of real-time polymerase chain reaction for quantification of microorganisms in environmental samples. Appl Microbiol Biotechnol 70:281–289
94. Smith CJ, Osborn AM (2009) Advantages and limitations of quantitative PCR (Q-PCR)-based approaches in microbial ecology. FEMS Microbiol Ecol 67:6–20
95. Gittel A, Sorensen KB, Skovhus TL, Ingvorsen K, Schramm A (2009) Prokaryotic community structure and sulfate reducer activity in water from high-temperature oil reservoirs with and without nitrate treatment. Appl Environ Microbiol 75:7086–7096
96. Leloup J, Loy A, Knab NJ, Borowski C, Wagner M, Jørgensen BB (2007) Diversity and abundance of sulfate-reducing microorganisms in the sulfate and methane zones of a marine sediment, Black Sea. Environ Microbiol 9:131–142
97. Schippers A, Kock D, Höft C, Köweker G, Siegert M (2012) Quantification of microbial communities in subsurface marine sediments of the Black Sea and off Namibia. Front Microbiol 3
98. Agrawal A, Lal B (2009) Rapid detection and quantification of bisulfite reductase genes in oil field samples using real-time PCR. FEMS Microbiol Ecol 69:301–312
99. Akkanen J, Tuikka A, Kukkonen JVK (2012) On the borderline of dissolved and particulate organic matter: Partitioning and bioavailability of polycyclic aromatic hydrocarbons. Ecotoxicol Environ Saf 78:91–98
100. Gittel A, Kofoed MVW, Sørensen KB, Ingvorsen K, Schramm A (2012) Succession of *Deferribacteres* and *Epsilonproteobacteria* through a nitrate-treated high-temperature oil production facility. Syst Appl Microbiol 35:165–174
101. Reis P, Waldron L, Goswami R, Xu W et al (2011) mRNA transcript quantification in archival samples using multiplexed, color-coded probes. BMC Biotechnol 11:46
102. Wang Q, Garrity GM, Tiedje JM, Cole JR (2007) Naïve Bayesian classifier for rapid assignment of rRNA sequences into the new bacterial taxonomy. Appl Environ Microbiol 73:5261–5267
103. Meyerdierks A, Kube M, Kostadinov I, Teeling H et al (2010) Metagenome and mRNA expression analyses of anaerobic methanotrophic archaea of the ANME-1 group. Environ Microbiol 12:422–439
104. Siegert M, Taubert M, Seifert J, von Bergen M, et al (in press) The nitrogen cycle in anaerobic methanotrophic mats of the Black Sea is linked to sulfate reduction and biomass decomposition. FEMS Microbiol Ecol. doi:10.1111/1574-6941.12156 (in press)

105. Metzker ML (2010) Sequencing technologies—the next generation. Nat Rev Genet 11:31–46
106. DeSantis T, Brodie E, Moberg J, Zubieta I, Piceno Y, Andersen G (2007) High-density universal 16S rRNA microarray analysis reveals broader diversity than typical clone library when sampling the environment. Microb Ecol 53:371–383
107. He Z, Deng Y, Van Nostrand JD, Tu Q et al (2010) GeoChip 3.0 as a high-throughput tool for analyzing microbial community composition, structure and functional activity. ISME J 4:1167–1179
108. Muyzer G, Smalla K (1998) Application of denaturing gradient gel electrophoresis (DGGE) and temperature gradient gel electrophoresis (TGGE) in microbial ecology. Antonie Leeuwenhoek 73:127–141
109. Liu WT, Marsh TL, Cheng H, Forney LJ (1997) Characterization of microbial diversity by determining terminal restriction fragment length polymorphisms of genes encoding 16S rRNA. Appl Environ Microbiol 63:4516–4522
110. Van Der Kraan GM, Bruining J, Lomans BP, Van Loosdrecht MCM, Muyzer G (2010) Microbial diversity of an oil–water processing site and its associated oil field: the possible role of microorganisms as information carriers from oil-associated environments. FEMS Microbiol Ecol 71:428–443
111. Radajewski S, Ineson P, Parekh NR, Murrell JC (2000) Stable-isotope probing as a tool in microbial ecology. Nature 403:646–649
112. Herrmann S, Kleinsteuber S, Chatzinotas A, Kuppardt S et al (2010) Functional characterization of an anaerobic benzene-degrading enrichment culture by DNA stable isotope probing. Environ Microbiol 12:401–411
113. Winderl C, Penning H, Netzer FV, Meckenstock RU, Lueders T (2010) DNA-SIP identifies sulfate-reducing Clostridia as important toluene degraders in tar-oil-contaminated aquifer sediment. ISME J 4:1314–1325
114. Redmond MC, Valentine DL, Sessions AL (2010) Identification of novel methane-, ethane-, and propane-oxidizing bacteria at marine hydrocarbon seeps by stable isotope probing. Appl Environ Microbiol 76:6412–6422
115. Wagner M (2009) Single-cell ecophysiology of microbes as revealed by Raman microspectroscopy or secondary ion mass spectrometry imaging. Annu Rev Microbiol 63:411–429
116. Hallmann C, Schwark L, Grice K (2008) Community dynamics of anaerobic bacteria in deep petroleum reservoirs. Nat Geosci 1:588–591
117. Bennett B, Adams JJ, Gray ND, Sherry A et al (2013) The controls on the composition of biodegraded oils in the deep subsurface. Part 3. The impact of microorganism distribution on petroleum geochemical gradients in biodegraded petroleum reservoirs. Org Geochem 56:94–105
118. Sadasivam S, Manickam A (1996) Nucleic acids. Biochemical methods. New Age International (P) Limited, Publishers, New Delhi
119. Jiménez N, Morris BEL, Cai M, Gründger F et al (2012) Evidence for in situ methanogenic oil degradation in the Dagang oil field. Org Geochem 52:44–54
120. Yamane K, Maki H, Nakayama T, Nakajima T et al (2008) Diversity and similarity of microbial communities in petroleum crude oils produced in asia. Biosci Biotechnol Biochem 72:2831–2839
121. Lappé M, Kallmeyer J (2011) A cell extraction method for oily sediments. Front Microbiol 2
122. Korenblum E, Souza DB, Penna M, Seldin L (2012) Molecular Analysis of the bacterial communities in crude oil samples from two Brazilian offshore petroleum platforms. Int J Microbiol 2012: Article ID 156537
123. Von Der Weid I, Korenblum E, Jurelevicius D, Rosado AS et al (2008) Molecular diversity of bacterial communities from subseafloor rock samples in a deep-water production basin in Brazil. J Microbiol Biotechnol 18:5–14
124. Kryachko Y, Dong X, Sensen C, Voordouw G (2012) Compositions of microbial communities associated with oil and water in a mesothermic oil field. Antonie Leeuwenhoek 101:493–506

125. Penner TJ, Foght JM, Budwill K (2010) Microbial diversity of western Canadian subsurface coal beds and methanogenic coal enrichment cultures. Int J Coal Geol 82:81–93
126. Klein DA, Flores RM, Venot C, Gabbert K et al (2008) Molecular sequences derived from Paleocene Fort Union Formation coals vs. associated produced waters: Implications for CBM regeneration. Int J Coal Geol 76:3–13
127. Karpicz R, Dementjev A, Kuprionis Z, Pakalnis S et al (2005) Laser fluorosensor for oil spot detection. Lith J Phys 45:213–218
128. Moter A, Göbel UB (2000) Fluorescence in situ hybridization (FISH) for direct visualization of microorganisms. J Microbiol Methods 41:85–112
129. U.S. Department of Energy, U.S. Environmental Protection Agency (2000) Carbon dioxide emissions from the generation of electric power in the United States. U.S. Department of Energy & U.S. Environmental Protection Agency, Washington, DC, USA
130. Head IM, Larter SR, Gray ND, Sherry A et al (2010) Hydrocarbon degradation in petroleum reservoirs. In: Timmis KN (ed) Handbook of hydrocarbon and lipid microbiology. Springer, Berlin
131. Tausson WO, Vesselov IJ (1934) On the bacteriology of the decomposition of cyclical compounds at the reduction of sulphates. Mikrobiologiia 3:360–369
132. Novelli GD, ZoBell CE (1944) Assimilation of petroleum hydrocarbons by sulfate-reducing bacteria. J Bacteriol 47:447–448
133. Muller F (1957) On methane fermentation of higher alkanes. Antonie Leeuwenhoek 23:369–384
134. Davis JB, Yarbrough HF (1966) Anaerobic oxidation of hydrocarbons by *Desulfovibrio desulfuricans*. Chem Geol 1:137–144
135. Grbić-Galić D, Vogel TM (1987) Transformation of toluene and benzene by mixed methanogenic cultures. Appl Environ Microbiol 53:254–260
136. Zengler K, Heider J, Rosselló-Mora R, Widdel F (1999) Phototrophic utilization of toluene under anoxic conditions by a new strain of *Blastochloris sulfoviridis*. Arch Microbiol 172:204–212
137. Dolfing J, Larter SR, Head IM (2007) Thermodynamic constraints on methanogenic crude oil biodegradation. ISME J 2:442–452
138. Mayumi D, Mochimaru H, Yoshioka H, Sakata S et al (2011) Evidence for syntrophic acetate oxidation coupled to hydrogenotrophic methanogenesis in the high-temperature petroleum reservoir of Yabase oil field (Japan). Environ Microbiol 13:1995–2006
139. Zengler K, Richnow HH, Rossello-Mora R, Michaelis W, Widdel F (1999) Methane formation from long-chain alkanes by anaerobic microorganisms. Nature 401:266–269
140. Gieg LM, Duncan KE, Suflita JM (2008) Bioenergy production via microbial conversion of residual oil to natural gas. Appl Environ Microbiol 74:3022–3029
141. Siegert M, Cichocka D, Herrmann S, Gründger F et al (2011) Accelerated methanogenesis from aliphatic and aromatic hydrocarbons under iron and sulfate reducing conditions. FEMS Microbiol Lett 315:6–16
142. Eberlein C, Johannes J, Mouttaki H, Sadeghi M et al (2013) ATP-dependent/-independent enzymatic ring reductions involved in the anaerobic catabolism of naphthalene. Environ Microbiol 15:1832–1841
143. Krüger M, Siegert M, Feisthauer S, Richnow H-H (2011) Microbial conversion of higher hydrocarbons to methane in oil and coal reservoirs. In: International symposium on applied microbiology and molecular biology in oil systems. Biodegradation Society, Calgary
144. Gray MR, Yeung A, Foght JM, Yarranton HW (2008) Potential microbial enhanced oil recovery processes: a critical analysis. In: SPE annual technical conference and exhibition. SPE Publications, Denver
145. Farias MJ, Watson RW (2007) Interaction of nitrogen/CO_2 mixtures with crude oil. Pennsylvania State University, University Park, PA, USA
146. Haynes Jr S, Alston RB (1990) Study of the mechanisms of carbon dioxide flooding and applications to more efficient EOR projects. In: SPE/DOE enhanced oil recovery symposium, 1990 Copyright 1990. Society of Petroleum Engineers Inc., Tulsa

147. Cramer SD (1984) Solubility of methane in brines from 0 to 300°C. Ind Eng Chem Process Design Dev 23:533–538
148. Michels A, Gerver J, Bijl A (1936) The influence of pressure on the solubility of gases. Physica 3:797–808
149. Pomeroy RD, Lacey WN, Scudder NF, Stapp FP (1933) Rate of solution of methane in quiescent liquid hydrocarbons. Ind Eng Chem 25:1014–1019
150. Gunsalus RP, Romesser JA, Wolfe RS (1978) Preparation of coenzyme M analogs and their activity in the methyl coenzyme M reductase system of *Methanobacterium thermoautotrophicum*. Biochemistry (Mosc) 17:2374–2377
151. Belyaev SS, Borzenkov IA, Nazina TN, Rozanova EP et al (2004) Use of microorganisms in the biotechnology for the enhancement of oil recovery. Microbiology 73:590–598
152. Updegraff DM (1957) Recovery of petroleum oil. United States, US Patent 2807570.
153. Hitzman DO, Sperl GT (1994) A new microbial technology for enhanced oil recovery and sulfide prevention and reduction. In: SPE/DOE improved oil recovery symposium. SPE Publications, Tulsa
154. Taber JJ (1969) Dynamic and static forces required to remove a discontinuous oil phase from porous media containing both oil and water. SPE J 9
155. Reed RL, Healy RN (1977) Some physicochemical aspects of microemulsion flooding: a review. In: Shah DO, Schecheter RS (eds) Improved oil recovery by surfactant and polymer flooding. Academic Press, New York
156. Finnerty WR, Singer ME (1984) A microbial biosurfactant–physiology, biochemistry, and applications. In: Nash CH (ed) Development in Industrial Microbiology. Society for Industrial Microbiology
157. Klein B, Bouriat P, Goulas P, Grimaud R (2009) Behavior of *Marinobacter hydrocarbonoclasticus* SP17 cells during initiation of biofilm formation at the alkane–water interface. Biotechnol Bioeng 105:461–468
158. Ivanov IB, Danov KD, Ananthapadmanabhan KP, Lips A (2005) Interfacial rheology of adsorbed layers with surface reaction: on the origin of the dilatational surface viscosity. Adv Colloid Interface Sci 114–115:61–92
159. Bouriat P, El Kerri N, Graciaa A, Lachaise J (2004) Properties of a two-dimensional asphaltene network at the water-cyclohexane interface deduced from dynamic tensiometry. Langmuir 20:7459–7464
160. Anton N, Vandamme TF, Bouriat P (2013) Dilatational rheology of a gel point network formed by nonionic soluble surfactants at the oil-water interface. Soft Matter 9:1310–1318
161. Rosenberg M (2006) Microbial adhesion to hydrocarbons: twenty-five years of doing MATH. FEMS Microbiol Lett 262:129–134
162. Weizmann C (1919) Production of acetone and alcohol by bacteriological processes. United States, US Patent 1315585
163. Neu TR (1996) Significance of bacterial surface-active compounds in interaction of bacteria with interfaces. Microbiol Rev 60:151–166
164. Rosenberg E (2006) Biosurfactants. In: Dworkin M et al (eds) The prokaryotes. Springer, New York
165. Riser-Roberts E (1998) 3.4 Intermediate metabolites and end products of biodegradation. Remediation of petroleum contaminated soils: biological, physical, and chemical processes. Lewis Publishers, London
166. Banat IM, Makkar RS, Cameotra SS (2000) Potential commercial applications of microbial surfactants. Appl Microbiol Biotechnol 53:495–508
167. Mulligan CN (2005) Environmental applications for biosurfactants. Environ Pollut 133:183–198
168. McInerney MJ, Javaheri M, Nagle DP (1990) Properties of the biosurfactant produced by *Bacillus licheniformis* strain JF-2. J Ind Microbiol Biotechnol 5:95–101
169. Amani H, Sarrafzadeh MH, Haghighi M, Mehrnia MR (2010) Comparative study of biosurfactant producing bacteria in MEOR applications. J Petrol Sci Eng 75:209–214

170. Deng D, Li C, Ju Q, Wu P, Dietrich FL, Zhou ZH (1999) Systematic extensive laboratory studies of microbial EOR mechanisms and microbial EOR application results in Changqing oilfield. In: SPE Asia Pacific oil and gas conference and exhibition. SPE Publications, Jakarta
171. Chowdhury SR, Manna S, Saha P, Basak RK et al (2011) Composition analysis and material characterization of an emulsifying extracellular polysaccharide (EPS) produced by *Bacillus megaterium* RB-05: a hydrodynamic sediment-attached isolate of freshwater origin. J Appl Microbiol 111:1381–1393
172. Zosim Z, Gutnick D, Rosenberg E (1982) Properties of hydrocarbon-in-water emulsions stabilized by *Acinetobacter* RAG-1 emulsan. Biotechnol Bioeng 24:281–292
173. Zhang Y, Miller R (1995) Effect of rhamnolipid (biosurfactant) structure on solubilization and biodegradation of *n*-alkanes. Appl Environ Microbiol 61:2247–2251
174. Löser C, Seidel H, Zehnsdorf A, Hoffmann P (2000) Improvement of the bioavailability of hydrocarbons by applying nonionic surfactants during the microbial remediation of a sandy soil. Acta Biotechnol 20:99–118
175. Pacheco GJ, Ciapina EMP, Gomes EdB, Pereira Junior N (2010) Biosurfactant production by *Rhodococcus erythropolis* and its application to oil removal. Braz J Microbiol 41: 685–693
176. Thavasi R, Jayalakshmi S, Banat IM (2011) Application of biosurfactant produced from peanut oil cake by *Lactobacillus delbrueckii* in biodegradation of crude oil. Bioresour Technol 102:3366–3372
177. Darvishi P, Ayatollahi S, Mowla D, Niazi A (2011) Biosurfactant production under extreme environmental conditions by an efficient microbial consortium, ERCPPI-2. Colloids Surf, B 84:292–300
178. Owsianiak M, Chrzanowski L, Szulc A, Staniewski J et al (2009) Biodegradation of diesel/biodiesel blends by a consortium of hydrocarbon degraders: Effect of the type of blend and the addition of biosurfactants. Bioresour Technol 100:1497–1500
179. Kunapuli U, Griebler C, Beller HR, Meckenstock RU (2008) Identification of intermediates formed during anaerobic benzene degradation by an iron-reducing enrichment culture. Environ Microbiol 10:1703–1712
180. Selesi D, Meckenstock RU (2009) Anaerobic degradation of the aromatic hydrocarbon biphenyl by a sulfate-reducing enrichment culture. FEMS Microbiol Ecol 68:86–93
181. Bogan BW, Sullivan WR, Kayser KJ, Derr KD, Aldrich HC, Paterek JR (2003) *Alkanindiges illinoisensis* gen. nov., sp. nov., an obligately hydrocarbonoclastic, aerobic squalane-degrading bacterium isolated from oilfield soils. Int J Syst Evol Microbiol 53:1389–1395
182. Golyshin PN, Chernikova TN, Abraham W-R, Lünsdorf H, Timmis KN, Yakimov MM (2002) *Oleiphilaceae* fam. nov., to include *Oleiphilus messinensis* gen. nov., sp. nov., a novel marine bacterium that obligately utilizes hydrocarbons. Int J Syst Evol Microbiol 52:901–911
183. Merkel GJ, Stapleton SS, Perry JJ (1978) Isolation and peptidoglycan of Gram-negative hydrocarbon-utilizing thermophilic bacteria. J Gen Microbiol 109:141–148
184. Radwan SS, Barabás G, Sorkhoh NA, Damjanovich S et al (1998) Hydrocarbon uptake by *Streptomyces*. FEMS Microbiol Lett 169:87–94
185. Yakimov MM, Giuliano L, Denaro R, Crisafi E et al (2004) *Thalassolituus oleivorans* gen. nov., sp. nov., a novel marine bacterium that obligately utilizes hydrocarbons. Int J Syst Evol Microbiol 54:141–148
186. Phillips WE, Perry JJ (1976) *Thermomicrobium fosteri* sp. nov., a hydrocarbon-utilizing obligate thermophile. Int J Syst Bacteriol 26:220–225
187. Feisthauer S, Siegert M, Seidel M, Richnow HH et al (2010) Isotopic fingerprinting of methane and CO_2 formation from aliphatic and aromatic hydrocarbons. Org Geochem 41:482–490
188. Kniemeyer O, Musat F, Sievert SM, Knittel K et al (2007) Anaerobic oxidation of short-chain hydrocarbons by marine sulphate-reducing bacteria. Nature 449:898–901

189. So CM, Phelps CD, Young LY (2003) Anaerobic transformation of alkanes to fatty acids by a sulfate-reducing bacterium, strain Hxd3. Appl Environ Microbiol 69:3892–3900
190. So CM, Young LY (1999) Initial reactions in anaerobic alkane degradation by a sulfate reducer, strain AK-01. Appl Environ Microbiol 65:5532–5540
191. Shaw J (1984) Enhanced oil recovery using carboxylate surfactant systems. J Am Oil Chemists' Soc 61:1389–1394
192. Jehmlich N, Schmidt F, von Bergen M, Richnow HH, Vogt C (2008) Protein-based stable isotope probing (protein-SIP) reveals active species within anoxic mixed cultures. ISME J 2:1122–1133
193. Kasai Y, Takahata Y, Manefield M, Watanabe K (2006) RNA-based stable isotope probing and isolation of anaerobic benzene-degrading bacteria from gasoline-contaminated groundwater. Appl Environ Microbiol 72:3586–3592
194. Meckenstock RU, Annweiler E, Michaelis W, Richnow HH, Schink B (2000) Anaerobic naphthalene degradation by a sulfate-reducing enrichment culture. Appl Environ Microbiol 66:2743–2747
195. Musat F, Galushko A, Jacob J, Widdel F et al (2009) Anaerobic degradation of naphthalene and 2-methylnaphthalene by strains of marine sulfate-reducing bacteria. Environ Microbiol 11:209–219
196. Johnsen AR, Winding A, Karlson U, Roslev P (2002) Linking of microorganisms to phenanthrene metabolism in soil by analysis of ^{13}C-labeled cell lipids. Appl Environ Microbiol 68:6106–6113
197. Rabus R (2005) Biodegradation of hydrocarbons under anoxic conditions. In: Ollivier B, Magot M (eds) Petroleum microbiology. ASM Press, Washington, DC
198. Shaw J.E., (1985) Immiscible displacement of oil with surfactant system. United States, US Patent 4556495.
199. Burland SM, Edwards EA (1999) Anaerobic benzene biodegradation linked to nitrate reduction. Appl Environ Microbiol 65:529–533
200. Harvey RR (1967) Oil recovery in water flooding. United States, US Patent 3326286
201. Cooper DG, Macdonald CR, Duff SJB, Kosaric N (1981) Enhanced production of surfactin from *Bacillus subtilis* by continuous product removal and metal cation additions. Appl Environ Microbiol 42:408–412
202. Javaheri M, Jenneman GE, McInerney MJ, Knapp RM (1985) Anaerobic production of a biosurfactant by *Bacillus licheniformis* JF-2. Appl Environ Microbiol 50:698–700
203. Maudgalya S, McInerney MJ, Knapp RM, Nagle DP, Folmsbee MJ (2004) Development of bio-surfactant based microbial enhanced oil recovery procedure. In: SPE/DOE symposium on improved oil recovery. SPE Publications, Tulsa
204. Maudgalya S, Knapp RM, McInerney MJ, Folmsbee M, Nagle DP (2005) Tertiary oil recovery with microbial biosurfactant treatment of low-permeability Berea sandstone cores. In: SPE production operations symposium. SPE Publications, Oklahoma City
205. Kim H-S, Yoon B-D, Lee C-H, Suh H–H et al (1997) Production and properties of a lipopeptide biosurfactant from *Bacillus subtilis* C9. J Ferment Bioeng 84:41–46
206. Zekri AY, Al-Khanbashi (2000) A microbial phase behavior laboratory studies. In: Abu Dhabi international petroleum exhibition and conference. SPE Publications, Abu Dhabi
207. Lin SC, Minton MA, Sharma MM, Georgiou G (1994) Structural and immunological characterization of a biosurfactant produced by *Bacillus licheniformis* JF-2. Appl Environ Microbiol 60:31–38
208. Nguyen TT, Youssef NH, McInerney MJ, Sabatini DA (2008) Rhamnolipid biosurfactant mixtures for environmental remediation. Water Res 42:1735–1743
209. Zekri AY, Almehaideb RA, Chaalal O (1999) Project of increasing oil recovery from UAE reservoirs using bacteria flooding. In: SPE annual technical conference and exhibition. SPE Publications, Houston
210. Kim H-S, Kim S-B, Park S-H, Oh H-M et al (2000) Expression of *sfp* gene and hydrocarbon degradation by *Bacillus subtilis*. Biotechnol Lett 22:1431–1436

211. Lee Y-K, Yoon B-D, Yoon J-H, Lee S-G et al (2007) Cloning of srfA operon from *Bacillus subtilis* C9 and its expression in *E. coli*. Appl Microbiol Biotechnol 75:567–572
212. Lin SC, Carswell KS, Sharma MM, Georgiou G (1994) Continuous production of the lipopeptide biosurfactant of *Bacillus licheniformis* JF-2. Appl Microbiol Biotechnol 41:281–285
213. Lin S-C, Lin K-G, Lo C-C, Lin Y-M (1998) Enhanced biosurfactant production by a *Bacillus licheniformis* mutant. Enzyme Microb Technol 23:267–273
214. Yakimov M, Timmis K, Wray V, Fredrickson H (1995) Characterization of a new lipopeptide surfactant produced by thermotolerant and halotolerant subsurface *Bacillus licheniformis* BAS50. Appl Environ Microbiol 61:1706–1713
215. Kang Z, Yeung A, Foght JM, Gray MR (2008) Mechanical properties of hexadecane-water interfaces with adsorbed hydrophobic bacteria. Colloids Surf B: Biointerfaces 62:273–279
216. Kang Z, Yeung A, Foght JM, Gray MR (2008) Hydrophobic bacteria at the hexadecane-water interface: examination of micrometre-scale interfacial properties. Colloids Surf B: Biointerfaces 67:59–66
217. Town K, Sheehy AJ, Govreau BR (2010) MEOR success in Southern Saskatchewan. SPE Reservoir Eval Eng 13
218. Illias RM, Ooi SW, Idris AK (1999) Isolation and characterization of thermophilic microorganisms from Malaysian oil fields. In: SPE annual technical conference and exhibition. SPE Publications, Houston
219. Halim AY, Fauzi UD, Siregar S, Soewono E et al (2009) Microbial enhanced oil recovery: an investigation of bacteria ability to live and alter crude oil physical characteristics in high pressure condition. In: Asia Pacific oil and gas conference & exhibition. SPE Publications, Jakarta
220. Kotlar HK, Wentzel A, Throne-Holst M, Zotchev S, Ellingsen T (2007) Wax control by biocatalytic degradation in high-paraffinic crude oils. In: International symposium on oilfield chemistry. SPE Publications, Houston
221. Maure MA, Saldaña AA, Juarez AR (2005) Biotechnology applications to EOR in Talara offshore oil fields, Northwest Peru. In: SPE Latin American and Caribbean petroleum engineering conference. SPE Publications, Rio de Janeiro
222. Ladygina N, Dedyukhina EG, Vainshtein MB (2006) A review on microbial synthesis of hydrocarbons. Process Biochem 41:1001–1014
223. Oremland RS (1981) Microbial formation of ethane in anoxic estuarine sediments. Appl Environ Microbiol 42:122–129
224. Belay N, Daniels L (1988) Ethane production by *Methanosarcina barkeri* during growth in ethanol supplemented medium. Antonie Leeuwenhoek 54:113–125
225. Hinrichs K-U, Hayes JM, Bach W, Spivack AJ et al (2006) Biological formation of ethane and propane in the deep marine subsurface. Proc Natl Acad Sci USA 103:14684–14689
226. Harder J, Foss S (1999) Anaerobic formation of the aromatic hydrocarbon *p*-cymene from monoterpenes by methanogenic enrichment cultures. Geomicrobiol J 16:295–305
227. Rude MA, Baron TS, Brubaker S, Alibhai M, Del Cardayre SB, Schirmer A (2011) Terminal olefin (1-alkene) biosynthesis by a novel P450 fatty acid decarboxylase from *Jeotgalicoccus* species. Appl Environ Microbiol 77:1718–1727
228. Sukovich DJ, Seffernick JL, Richman JE, Gralnick JA, Wackett LP (2010) Widespread head-to-head hydrocarbon biosynthesis in bacteria and role of OleA. Appl Environ Microbiol 76:3850–3862
229. Price LC, Wenger LM, Ging T, Blount CW (1983) Solubility of crude oil in methane as a function of pressure and temperature. Org Geochem 4:201–221
230. Shiu WY, Bobra M, Bobra AM, Maijanen A, Suntio L, Mackay D (1990) The water solubility of crude oils and petroleum products. Oil Chem Pollut 7:57–84
231. Holm LW (1974) Miscible flooding process using methane-enriched soluble oil. United States, US Patent 3830301

232. Wawrik B, Mendivelso M, Parisi VA, Suflita JM et al (2012) Field and laboratory studies on the bioconversion of coal to methane in the San Juan Basin. FEMS Microbiol Ecol 81:26–42
233. Hitzman DO, Dennis M, Hitzman DC (2004) Recent successes: MEOR using synergistic H_2S prevention and increased oil recovery systems. In: SPE/DOE symposium on improved oil recovery. SPE Publications, Tulsa
234. Zhang T, Gannon SM, Nevin KP, Franks AE, Lovley DR (2010) Stimulating the anaerobic degradation of aromatic hydrocarbons in contaminated sediments by providing an electrode as the electron acceptor. Environ Microbiol 12:1011–1020
235. Thompson BG, Jack TR (1984) Method of enhancing oil recovery by use of exopolymer producing microorganisms. United States, US Patent 4460043
236. Kantzas A, Stehmeier L, Marentette DF, Ferris FG, Jha KN, Maurits FM (1992) A novel method of sand consolidation through bacteriogenic mineral plugging. In: Annual technical meeting, Calgary
237. Khan HA, Gbosi A, Britton LN, Bryant SL (2008) Mechanistic models of microbe growth in heterogeneous porous media. In: SPE/DOE symposium on improved oil recovery, Tulsa
238. Nagase K, Zhang ST, Asami H, Yazawa N et al (2001) Improvement of sweep efficiency by microbial EOR process in Fuyu oilfield, China. In: SPE Asia Pacific oil and gas conference and exhibition. SPE Publications, Jakarta
239. Jenneman GE, Knapp RM, McInerney MJ, Menzie DE, Revus DE (1984) Experimental studies of in-situ microbial enhanced oil recovery. SPE J 24
240. Sugai Y, Hong C, Chida T, Enomoto H (2007) Simulation studies on the mechanisms and performances of MEOR using a polymer producing microorganism *Clostridium* sp. TU-15A. in Asia Pacific oil and gas conference and exhibition. SPE Publications, Jakarta
241. Stiles WE (1949) Use of permeability distributions in waterflood calculations. Trans AIME 15:9–12
242. Nagase K, Zhang ST, Asami H, Yazawa N et al (2002) A successful field test of microbial EOR process in Fuyu oilfield, China. In: SPE/DOE improved oil recovery symposium. SPE Publications, Tulsa
243. Reksidler R, Volpon AG, Barbosa LCF, Brasileiro CG, Nicolau HCC, Calasans Júnior JA (2010) A microbial enhanced oil recovery field pilot in a Brazilian onshore oilfield. In: SPE improved oil recovery symposium. SPE Publications, Tulsa
244. Knapp RM, McInerney MJ, Coates JD, Chisholm JL, Menzie DE, Bhupathiraju VK, (1992) Design and implementation of a microbially enhanced oil recovery field pilot, Payne County, Oklahoma. In: SPE annual technical conference and exhibition. SPE Publications, Washington
245. Stocks-Fischer S, Galinat JK, Bang SS (1999) Microbiological precipitation of $CaCO_3$. Soil Biol Biochem 31:1563–1571
246. Dupraz C, Reid RP, Braissant O, Decho AW, Norman RS, Visscher PT (2009) Processes of carbonate precipitation in modern microbial mats. Earth Sci Rev 96:141–162
247. Glunk C, Dupraz C, Braissant O, Gallagher KL, Verrecchia EP, Visscher PT (2011) Microbially mediated carbonate precipitation in a hypersaline lake, Big Pond (Eleuthera, Bahamas). Sedimentology 58:720–736
248. Bosak T, Newman DK (2003) Microbial nucleation of calcium carbonate in the Precambrian. Geology 31:577–580
249. Drobner E, Huber H, Wachtershauser G, Rose D, Stetter KO (1990) Pyrite formation linked with hydrogen evolution under anaerobic conditions. Nature 346:742–744
250. Friedmann EI (1982) Endolithic microorganisms in the Antarctic Cold Desert. Science 215:1045–1053
251. Thorseth IH, Furnes H, Heldal M (1992) The importance of microbiological activity in the alteration of natural basaltic glass. Geochim Cosmochim Acta 56:845–850
252. Blackhurst RL, Genge MJ, Kearsley AT, Grady MM (2005) Cryptoendolithic alteration of Antarctic sandstones: Pioneers or opportunists? J Geophys Res 110:E12S24

253. Udegbunam EO, Adkins JP, Knapp RM, McInerney MJ, Tanner RS (1991) Assessing the effects of microbial metabolism and metabolites on reservoir pore structure. In: SPE annual technical conference and exhibition. SPE Publications, Dallas
254. Wagner M, Müller P (1980) Verfahren zur mikrobiellen Sekundärgewinnung von Erdöl. German Democratic Republic, Patent DD000000140067A1
255. Yarbrough HF, Coty VF (1983) Microbially enhanced oil recovery from the upper cretaceous nacatoch formation, Union County, Arkansas. In: Proceedings of the 1982 international conference on microbial enhancement of oil Recovery (CONF-8205140), Bartlesville, USA
256. Burchfield TE, Carroll Jr. HB (1988) Economics of MEOR processes. In: Applications of microorganisms to petroleum technology, 12–13 August 1987. U.S. Department of Energy, Bartlesville, USA
257. Portwood JT (1995) A commercial microbial enhanced oil recovery technology: evaluation of 322 projects. In: SPE production operations symposium. SPE Publications, Oklahoma City
258. Maure A, Dietrich F, Gomez U, Vallesi J, Irusta M (2001) Waterflooding optimization using biotechnology: 2-year field test, La Ventana Field, Argentina. In: SPE Latin American and Caribbean petroleum engineering conference. SPE Publications, Buenos Aires
259. Maudgalya S, Knapp RM, McInerney MJ (2007) Microbially enhanced oil recovery technologies: a review of the past, present and future. In: Production and operations symposium. SPE Publications, Oklahoma City
260. Vadie AA, Stephens JO, Brown LR (1996) Utilization of indigenous microflora in permeability profile modification of oil bearing formations. In: SPE/DOE improved oil recovery symposium. SPE Publications, Tulsa
261. Hyne NJ (1991) Dictionary of petroleum exploration, drilling & production. PennWell Publishing Company, Tulsa
262. Ferguson KR, Lloyd CT, Dan S, James H (1996) Microbial pilot test for the control of paraffin and asphaltenes at Prudhoe Bay. In: SPE annual technical conference and exhibition. SPE Publications, Denver
263. Maure MA, Dietrich FL, Diaz VA, Argañaraz H (1999) Microbial enhanced oil recovery pilot test in Piedras Coloradas Field, Argentina. In: Latin American and Caribbean petroleum engineering conference. SPE Publications, Caracas, pp. 1–28
264. Talukdar S, Instefjord R (2008) Reservoir management of the Gullfaks main field. In: Europec/EAGE conference and exhibition. SPE Publications, Rome
265. Zhang X, Xiang T (2010) Review on microbial enhanced oil recovery technology and development in China. Int J Petrol Sci Technol 4:61–80
266. Gao CH, Zekri A (2011) Applications of microbial-enhanced oil recovery technology in the past decade. Energy Sources Part A: Recov Util Environ Effects 33:972–989
267. Mnif S, Chamkha M, Labat M, Sayadi S (2011) Simultaneous hydrocarbon biodegradation and biosurfactant production by oilfield-selected bacteria. J Appl Microbiol 111:525–536
268. Rees GN, Patel BKC, Grassia GS, Sheehy AJ (1997) *Anaerobaculum thermoterrenum* gen. nov., sp. nov., a novel, thermophilic bacterium which ferments citrate. Int J Syst Bacteriol 47:150–154
269. Kato T, Haruki M, Imanaka T, Morikawa M, Kanaya S (2001) Isolation and characterization of long-chain-alkane degrading *Bacillus thermoleovorans* from deep subterranean petroleum reservoirs. J Biosci Bioeng 91:64–70
270. Fardeau ML, Magot M, Patel BKC, Thomas P, Garcia JL, Ollivier B (2000) *Thermoanaerobacter subterraneus* sp. nov., a novel thermophile isolated from oilfield water. Int J Syst Evol Microbiol 50:2141–2149
271. Fardeau M-L, Salinas MB, L'Haridon S, Jeanthon C et al (2004) Isolation from oil reservoirs of novel thermophilic anaerobes phylogenetically related to *Thermoanaerobacter subterraneus*: reassignment of *T. subterraneus*, *Thermoanaerobacter yonseiensis*, *Thermoanaerobacter tengcongensis* and *Carboxydibrachium pacificum* to *Caldanaerobacter subterraneus* gen. nov., sp. nov., comb. nov. as four novel subspecies. Int J Syst Evol Microbiol 54:467–474

272. Ying J-Y, Wang B-J, Yang S-S, Liu S-J (2006) *Cyclobacterium lianum* sp. nov., a marine bacterium isolated from sediment of an oilfield in the South China Sea, and emended description of the genus Cyclobacterium. Int J Syst Evol Microbiol 56:2927–2930
273. Greene AC, Patel BKC, Sheehy AJ (1997) *Deferribacter thermophilus* gen. nov., sp. nov., a novel thermophilic manganese- and iron-reducing bacterium isolated from a petroleum reservoir. Int J Syst Bacteriol 47:505–509
274. Rees GN, Grassia GS, Sheehy AJ, Dwivedi PP, Patel BKC (1995) *Desulfacinum infernum* gen. nov., sp. nov., a thermophilic sulfate-reducing bacterium from a petroleum reservoir. Int J Syst Bacteriol 45:85–89
275. Rozanova EP, Tourova TP, Kolganova TV, Lysenko AM et al (2001) *Desulfacinum subterraneum* sp. nov., a new thermophilic sulfate-reducing bacterium isolated from a high-temperature oil field. Microbiology (Moscow, Russ. Fed.) 70:466–471
276. Lien T, Beeder J (1997) *Desulfobacter vibrioformis* sp. nov., a sulfate reducer from a water-oil separation system. Int J Syst Bacteriol 47:1124–1128
277. Harms G, Zengler K, Rabus R, Aeckersberg F et al (1999) Anaerobic oxidation of *o*-xylene, *m*-xylene, and homologous alkylbenzenes by new types of sulfate-reducing bacteria. Appl Environ Microbiol 65:999–1004
278. Lien T, Madsen M, Steen IH, Gjerdevik K (1998) *Desulfobulbus rhabdoformis* sp. nov., a sulfate reducer from a water-oil separation system. Int J Syst Bacteriol 48:469–474
279. Aeckersberg F, Bak F, Widdel F (1991) Anaerobic oxidation of saturated hydrocarbons to CO_2 by a new type of sulfate-reducing bacterium. Arch Microbiol 156:5–14
280. Davidova IA, Duncan KE, Choi OK, Suflita JM (2006) *Desulfoglaeba alkanexedens* gen. nov., sp. nov., an *n*-alkane-degrading, sulfate-reducing bacterium. Int J Syst Evol Microbiol 56:2737–2742
281. Tardy-Jacquenod C, Magot M, Patel BKC, Matheron R, Caumette P (1998) *Desulfotomaculum halophilum* sp. nov., a halophilic sulfate-reducing bacterium isolated from oil production facilities. Int J Syst Bacteriol 48:333–338
282. Nilsen RK, Torsvik T, Lien T (1996) *Desulfotomaculum thermocisternum* sp. nov., a sulfate reducer isolated from a hot north sea oil reservoir. Int J Syst Bacteriol 46:397–402
283. Belyakova E, Rozanova E, Borzenkov I, Tourova T et al (2006) The new facultatively chemolithoautotrophic, moderately halophilic, sulfate-reducing bacterium *Desulfovermiculus halophilus* gen. nov., sp. nov., isolated from an oil field. Microbiology 75:161–171
284. Feio MJ, Zinkevich V, Beech IB, Llobet-Brossa E et al (2004) *Desulfovibrio alaskensis* sp. nov., a sulphate-reducing bacterium from a soured oil reservoir. Int J Syst Evol Microbiol 54:1747–1752
285. Magot M, Basso O, Tardy-Jacquenod C, Caumette P (2004) *Desulfovibrio bastinii* sp. nov. and *Desulfovibrio gracilis* sp. nov., moderately halophilic, sulfate-reducing bacteria isolated from deep subsurface oilfield water. Int J Syst Evol Microbiol 54:1693–1697
286. Miranda-Tello E, Fardeau M-L, Fernández L, Ramírez F et al (2003) *Desulfovibrio capillatus* sp. nov., a novel sulfate-reducing bacterium isolated from an oil field separator located in the Gulf of Mexico. Anaerobe 9:97–103
287. Tardy-Jacquenod C, Magot M, Laigret F, Kaghad M et al (1996) *Desulfovibrio gabonensis* sp. nov., a new moderately halophilic sulfate-reducing bacterium isolated from an oil pipeline. Int J Syst Bacteriol 46:710–715
288. Magot M, Caumette P, Desperrier JM, Matheron R et al (1992) *Desulfovibrio longus* sp. nov., a sulfate-reducing bacterium isolated from an oil-producing well. Int J Syst Bacteriol 42:398–402
289. Nga DP, Ha DTC, Hien LT, Stan-Lotter H (1996) *Desulfovibrio vietnamensis* sp.nov., a halophilic sulfate-reducing bacterium from vietnamese oil fields. Anaerobe 2:385–392
290. Magot M, Ravot G, Campaignolle X, Ollivier B et al (1997) *Dethiosulfovibrio peptidovorans* gen. nov., sp. nov., a new anaerobic, slightly halophilic, thiosulfate-reducing bacterium from corroding offshore oil wells. Int J Syst Bacteriol 47:818–824

291. Wang X-B, Chi C-Q, Nie Y, Tang Y-Q et al (2011) Degradation of petroleum hydrocarbons (C6-C40) and crude oil by a novel *Dietzia* strain. Bioresour Technol 102:7755–7761
292. Ravot G, Magot M, Fardeau M-L, Patel BKC et al (1999) *Fusibacter paucivorans* gen. nov., sp. nov., an anaerobic, thiosulfate-reducing bacterium from an oil-producing well. Int J Syst Bacteriol 49:1141–1147
293. Miranda-Tello E, Fardeau M-L, Sepulveda J, Fernandez L et al (2003) *Garciella nitratireducens* gen. nov., sp. nov., an anaerobic, thermophilic, nitrate- and thiosulfate-reducing bacterium isolated from an oilfield separator in the Gulf of Mexico. Int J Syst Evol Microbiol 53:1509–1514
294. Greene AC, Patel BKC, Yacob S (2009) *Geoalkalibacter subterraneus* sp. nov., an anaerobic Fe(III)- and Mn(IV)-reducing bacterium from a petroleum reservoir, and emended descriptions of the family *Desulfuromonadaceae* and the genus *Geoalkalibacter*. Int J Syst Evol Microbiol 59:781–785
295. Nazina TN, Sokolova DS, Grigoryan AA, Shestakova NM et al (2005) *Geobacillus jurassicus* sp. nov., a new thermophilic bacterium isolated from a high-temperature petroleum reservoir, and the validation of the *Geobacillus* species. Syst Appl Microbiol 28:43–53
296. Nazina TN, Tourova TP, Poltaraus AB, Novikova EV et al (2001) Taxonomic study of aerobic thermophilic bacilli: descriptions of *Geobacillus subterraneus* gen. nov., sp. nov. and *Geobacillus uzenensis* sp. nov. from petroleum reservoirs and transfer of *Bacillus stearothermophilus*, *Bacillus thermocatenulatus*, *Bacillus thermoleovorans*, *Bacillus kaustophilus*, *Bacillus thermoglucosidasius* and *Bacillus thermodenitrificans* to *Geobacillus* as the new combinations *G. stearothermophilus*, *G. thermocatenulatus*, *G. thermoleovorans*, *G. kaustophilus*, *G. thermoglucosidasius* and *G. thermodenitrificans*. Int J Syst Evol Microbiol 51:433–446
297. Kuisiene N, Raugalas J, Chitavichius D (2004) *Geobacillus lituanicus* sp. nov. Int J Syst Evol Microbiol 54:1991–1995
298. Zheng C, He J, Wang Y, Wang M, Huang Z (2011) Hydrocarbon degradation and bioemulsifier production by thermophilic *Geobacillus pallidus* strains. Bioresour Technol 102:9155–9161
299. Zhang J, Zhang X, Liu J, Li R, Shen B (2012) Isolation of a thermophilic bacterium, *Geobacillus* sp. SH-1, capable of degrading aliphatic hydrocarbons and naphthalene simultaneously, and identification of its naphthalene degrading pathway. Bioresour Technol 124:83–89
300. Xue Y, Sun X, Zhou P, Liu R, Liang F, Ma Y (2003) *Gordonia paraffinivorans* sp. nov., a hydrocarbon-degrading actinomycete isolated from an oil-producing well. Int J Syst Evol Microbiol 53:1643–1646
301. Ravot G, Magot M, Ollivier B, Patel BKC et al (1997) *Haloanaerobium congolense* sp. nov., an anaerobic, moderately halophilic, thiosulfate- and sulfur-reducing bacterium from an African oil field. FEMS Microbiol Lett 147:81–88
302. DiPippo JL, Nesbø CL, Dahle H, Doolittle WF, Birkland N-K, Noll KM (2009) *Kosmotoga olearia* gen. nov., sp. nov., a thermophilic, anaerobic heterotroph isolated from an oil production fluid. Int J Syst Evol Microbiol 59:2991–3000
303. Salinas MB, Fardeau M-L, Thomas P, Cayol J-L, Patel BKC, Ollivier B (2004) *Mahella australiensis* gen. nov., sp. nov., a moderately thermophilic anaerobic bacterium isolated from an Australian oil well. Int J Syst Evol Microbiol 54:2169–2173
304. Ollivier B, Fardeau M-L, Cayol J-L, Magot M et al (1998) *Methanocalculus halotolerans* gen. nov., sp. nov., isolated from an oil-producing well. Int J Syst Bacteriol 48:821–828
305. Nilsen RK, Torsvik T (1996) *Methanococcus thermolithotrophicus* isolated from North Sea oil field reservoir water. Appl Environ Microbiol 62:728–731
306. Cheng L, Qiu T-L, Yin X-B, Wu X-L et al (2007) *Methermicoccus shengliensis* gen. nov., sp. nov., a thermophilic, methylotrophic methanogen isolated from oil-production water, and proposal of *Methermicoccaceae* fam. nov. Int J Syst Evol Microbiol 57:2964–2969

307. Ben Ali Gam Z, Abdelkafi S, Casalot L, Tholozan JL, Oueslati R, Labat M (2007) *Modicisalibacter tunisiensis* gen. nov., sp. nov., an aerobic, moderately halophilic bacterium isolated from an oilfield-water injection sample, and emended description of the family *Halomonadaceae* Franzmann et al. 1989 emend Dobson and Franzmann 1996 emend. Ntougias et al. 2007. Int J Syst Evol Microbiol 57:2307–2313
308. Schippers A, Schumann P, Sproer C (2005) *Nocardioides oleivorans* sp. nov., a novel crude-oil-degrading bacterium. Int J Syst Evol Microbiol 55:1501–1504
309. Jayasinghearachchi HS, Lal B (2011) *Oceanotoga teriensis* gen. nov., sp. nov., a thermophilic bacterium isolated from offshore oil-producing wells. Int J Syst Evol Microbiol 61:554–560
310. Grabowski A, Tindall BJ, Bardin V, Blanchet D, Jeanthon C (2005) *Petrimonas sulfuriphila* gen. nov., sp. nov., a mesophilic fermentative bacterium isolated from a biodegraded oil reservoir. Int J Syst Evol Microbiol 55:1113–1121
311. Salinas MB, Fardeau M-L, Cayol J-L, Casalot L et al (2004) *Petrobacter succinatimandens* gen. nov., sp. nov., a moderately thermophilic, nitrate-reducing bacterium isolated from an Australian oil well. Int J Syst Evol Microbiol 54:645–649
312. Miranda-Tello E, Fardeau M-L, Joulian C, Magot M et al (2007) *Petrotoga halophila* sp. nov., a thermophilic, moderately halophilic, fermentative bacterium isolated from an offshore oil well in Congo. Int J Syst Evol Microbiol 57:40–44
313. Purwasena IA, Sugai Y, Sasaki K (2009) estimation of the potential of an oil-viscosity-reducing bacteria, *Petrotoga* isolated from an oilfield for MEOR. In: International petroleum technology conference. SPE Publications, Doha
314. Miranda-Tello E, Fardeau M-L, Thomas P, Ramirez F et al (2004) *Petrotoga mexicana* sp. nov., a novel thermophilic, anaerobic and xylanolytic bacterium isolated from an oil-producing well in the Gulf of Mexico. Int J Syst Evol Microbiol 54:169–174
315. Lien T, Madsen M, Rainey FA, Birkeland N-K (1998) *Petrotoga mobilis* sp. nov., from a North Sea oil-production well. Int J Syst Bacteriol 48:1007–1013
316. Samir MA, Ela MAE, Marsafy SE, El-Tayeb S, Sayyouh H, Wally AW (2010) New incubated *Pseudomonas aeruginosa* bacteria for increasing oil recovery under reservoir conditions. In: North Africa technical conference and exhibition, Cairo, Egypt
317. Zheng C, Yu L, Huang L, Xiu J, Huang Z (2012) Investigation of a hydrocarbon-degrading strain, *Rhodococcus ruber* Z25, for the potential of microbial enhanced oil recovery. J Petrol Sci Eng 81:49–56
318. Magot M, Fardeau M-L, Arnauld O, Lanau C et al (1997) *Spirochaeta smaragdinae* sp. nov., a new mesophilic strictly anaerobic spirochete from an oil field. FEMS Microbiol Lett 155:185–191
319. Cayol JL, Ollivier B, Patel BKC, Ravot G et al (1995) Description of *Thermoanaerobacter brockii* subsp. *lactiethylicus* subsp. nov., isolated from a deep subsurface French oil well, a proposal to reclassify *Thermoanaerobacter finnii* as *Thermoanaerobacter brockii* subsp. *finnii* comb. nov., and an emended description of *Thermoanaerobacter brockii*. Int J Syst Bacteriol 45:783–789
320. Feng Y, Cheng L, Zhang X, Li X, Deng Y, Zhang H (2010) *Thermococcoides shengliensis* gen. nov., sp. nov., a new member of the order *Thermotogales* isolated from oil-production fluid. Int J Syst Evol Microbiol 60:932–937
321. Beeder J, Torsvik T, Lien T (1995) *Thermodesulforhabdus norvegicus* gen. nov., sp. nov., a novel thermophilic sulfate-reducing bacterium from oil field water. Arch Microbiol 164:331–336
322. Ravot G, Magot M, Fardeau ML, Patel BKC et al (1995) *Thermotoga elfii* sp. nov., a novel thermophilic bacterium from an african oil-producing well. Int J Syst Bacteriol 45:308–314
323. Dahle H, Birkeland N-K (2006) *Thermovirga lienii* gen. nov., sp. nov., a novel moderately thermophilic, anaerobic, amino-acid-degrading bacterium isolated from a North Sea oil well. Int J Syst Evol Microbiol 56:1539–1545

324. Ying J-Y, Wang B-J, Dai X, Yang S-S, Liu S-J, Liu Z-P (2007) *Wenxinia marina* gen. nov., sp. nov., a novel member of the *Roseobacter* clade isolated from oilfield sediments of the South China Sea. Int J Syst Evol Microbiol 57:1711–1716
325. Dore SY, Clancy QE, Rylee SM, Kulpa CF (2003) Naphthalene-utilizing and mercury-resistant bacteria isolated from an acidic environment. Appl Microbiol Biotechnol 63:194–199
326. Meyer S, Moser R, Neef A, Stahl U, Kampfer P (1999) Differential detection of key enzymes of polyaromatic-hydrocarbon-degrading bacteria using PCR and gene probes. Microbiology 145:1731–1741
327. Chaîneau CH, Morel J, Dupont J, Bury E, Oudot J (1999) Comparison of the fuel oil biodegradation potential of hydrocarbon-assimilating microorganisms isolated from a temperate agricultural soil. Sci Total Environ 227:237–247
328. Amund OO, Higgins IJ (1985) The degradation of 1-phenylalkanes by an oil-degrading strain of *Acinetobacter lwoffi*. Antonie Leeuwenhoek 51:45–56
329. Mara K, Decorosi F, Viti C, Giovannetti L et al (2012) Molecular and phenotypic characterization of *Acinetobacter* strains able to degrade diesel fuel. Res Microbiol 163:161–172
330. Marín M, Pedregosa A, Laborda F (1996) Emulsifier production and microscopical study of emulsions and biofilms formed by the hydrocarbon-utilizing bacteria *Acinetobacter calcoaceticus* MM5. Appl Microbiol Biotechnol 44:660–667
331. Marín M, Pedregosa A, Ríos S, Laborda F (1996) Study of factors influencing the degradation of heating oil by *Acinetobacter calcoaceticus* MM5. Int Biodeterior Biodegrad 38:69–75
332. Büttner H (1926) Zur Kenntnis der Mykobakterien, insbesondere ihres quantitativen Stoffwechsels auf Paraffinnährböden. Archiv für Hygiene 97:12–27
333. Odokuma LO, Dickson AA (2003) Bioremediation of a crude oil polluted tropical rain forest soil. Global J Environ Sci 2:29–40
334. Prantera MT, Drozdowicz A, Gomes Leite S, Soares Rosado A (2002) Degradation of gasoline aromatic hydrocarbons by two N_2-fixing soil bacteria. Biotechnol Lett 24:85–89
335. Ridgway HF, Safarik J, Phipps D, Carl P, Clark D (1990) Identification and catabolic activity of well-derived gasoline-degrading bacteria from a contaminated aquifer. Appl Environ Microbiol 56:3565–3575
336. Yousaf S, Andria V, Reichenauer TG, Smalla K, Sessitsch A (2010) Phylogenetic and functional diversity of alkane degrading bacteria associated with Italian ryegrass (*Lolium multiflorum*) and Birdsfoot trefoil (*Lotus corniculatus*) in a petroleum oil-contaminated environment. J Hazard Mater 184:523–532
337. Yakimov MM, Golyshin PN, Lang S, Moore ERB et al (1998) *Alcanivorax borkumensis* gen. nov., sp. nov., a new, hydrocarbon-degrading and surfactant-producing marine bacterium. Int J Syst Bacteriol 48:339–348
338. Liu C, Shao Z (2005) *Alcanivorax dieselolei* sp. nov., a novel alkane-degrading bacterium isolated from sea water and deep-sea sediment. Int J Syst Evol Microbiol 55:1181–1186
339. Bruns A, Berthe-Corti L (1999) *Fundibacter jadensis* gen. nov., sp. nov., a new slightly halophilic bacterium, isolated from intertidal sediment. Int J Syst Bacteriol 49:441–448
340. Fernandez-Martinez J, Pujalte MJ, Garcia-Martinez J, Mata M, Garay E, Rodriguez-Valera F (2003) Description of *Alcanivorax venustensis* sp. nov. and reclassification of *Fundibacter jadensis* DSM 12178T (Bruns and Berthe-Corti 1999) as *Alcanivorax jadensis* comb. nov., members of the emended genus *Alcanivorax*. Int J Syst Evol Microbiol 53:331–338
341. Weelink SAB, Tan NCG, ten Broeke H, van den Kieboom C et al (2008) Isolation and characterization of *Alicycliphilus denitrificans* strain BC, which grows on benzene with chlorate as the electron acceptor. Appl Environ Microbiol 74:6672–6681
342. Iwabuchi N, Sunairi M, Urai M, Itoh C et al (2002) Extracellular polysaccharides of *Rhodococcus rhodochrous* S-2 stimulate the degradation of aromatic components in crude oil by indigenous marine bacteria. Appl Environ Microbiol 68:2337–2343

343. Khelifi N, Grossi V, Hamdi M, Dolla A et al (2010) Anaerobic oxidation of fatty acids and alkenes by the hyperthermophilic sulfate-reducing archaeon *Archaeoglobus fulgidus*. Appl Environ Microbiol 76:3057–3060
344. Rabus R, Jarling R, Lahme S, Kühner S et al (2011) Co-metabolic conversion of toluene in anaerobic *n*-alkane-degrading bacteria. Environ Microbiol 13:2576–2586
345. Rabus R, Widdel F (1995) Anaerobic degradation of ethylbenzene and other aromatic hydrocarbons by new denitrifying bacteria. Arch Microbiol 163:96–103
346. Song B, Haggblom MM, Zhou J, Tiedje JM, Palleroni NJ (1999) Taxonomic characterization of denitrifying bacteria that degrade aromatic compounds and description of *Azoarcus toluvorans* sp. nov. and *Azoarcus toluclasticus* sp. nov. Int J Syst Bacteriol 49:1129–1140
347. Fries MR, Forney LJ, Tiedje JM (1997) Phenol- and toluene-degrading microbial populations from an aquifer in which successful trichloroethene cometabolism occurred. Appl Environ Microbiol 63:1523–1530
348. Zhou J, Fries MR, Chee-Sanford JC, Tiedje JM (1995) Phylogenetic analyses of a new group of denitrifiers capable of anaerobic growth on toluene and description of *Azoarcus tolulyticus* sp. nov. Int J Syst Bacteriol 45:500–506
349. Gradova NB, Gornova IB, Eddaudi R, Salina RN (2003) Use of bacteria of the genus *Azotobacter* for bioremediation of oil-contaminated soils. Appl Biochem Microbiol 39:279–281
350. Christova N, Tuleva B, Nikolova-Damyanovab B (2004) Enhanced hydrocarbon biodegradation by a newly isolated *Bacillus subtilis* strain. Zeitschrift für Naturforschung 59c:205–208
351. Stucki G, Alexander M (1987) Role of dissolution rate and solubility in biodegradation of aromatic compounds. Appl Environ Microbiol 53:292–297
352. Atlas RM, Bartha R (1972) Degradation and mineralization of petroleum by two bacteria isolated from coastal waters. Biotechnol Bioeng 14:297–308
353. Pirnik MP, Atlas RM, Bartha R (1974) Hydrocarbon metabolism by *Brevibacterium erythrogenes*: normal and branched alkanes. J Bacteriol 119:868–878
354. Juhasz AL, Britz ML, Stanley GA (1997) Degradation of fluoranthene, pyrene, benz[a]anthracene and dibenz[a,h]anthracene by *Burkholderia cepacia*. J Appl Microbiol 83:189–198
355. Parales RE, Ditty JL, Harwood CS (2000) Toluene-degrading bacteria are chemotactic towards the environmental pollutants benzene, toluene, and trichloroethylene. Appl Environ Microbiol 66:4098–4104
356. Landuyt SM, Hsu EJ, Wang B, Tsay S (1995) Conversion of paraffin oil to alcohols by *Clostridium thermosaccharolyticum*. Appl Environ Microbiol 61:1153–1155
357. Zajic JE, Guignard H, Gerson DF (1977) Emulsifying and surface active agents from *Corynebacterium hydrocarboclastus*. Biotechnol Bioeng 19:1285–1301
358. Maure A, Dietrich F, Gomez U, Vallesi J, Irusta M (2001) Waterflooding Optimization Using Biotechnology: 2-Year Field Test, La Ventana Field, Argentina, SPE Latin American and Caribbean Petroleum Engineering Conference, SPE Publications, Buenos Aires, Argentina. SPE accession number 69652, doi: 10.2118/69652-MS
359. Dyksterhouse SE, Gray JP, Herwig RP, Lara JC, Staley JT (1995) *Cycloclasticus pugetii* gen. nov., sp. nov., an aromatic hydrocarbon-degrading bacterium from marine sediments. Int J Syst Bacteriol 45:116–123
360. Chung WK, King GM (2001) Isolation, characterization, and polyaromatic hydrocarbon degradation potential of aerobic bacteria from marine macrofaunal burrow sediments and description of *Lutibacterium anuloederans* gen. nov., sp. nov., and *Cycloclasticus spirillensus* sp. nov. Appl Environ Microbiol 67:5585–5592
361. Coates JD, Chakraborty R, Lack JG, O'Connor SM et al (2001) Anaerobic benzene oxidation coupled to nitrate reduction in pure culture by two strains of *Dechloromonas*. Nature 411:1039–1043

362. Salinero K, Keller K, Feil W, Feil H et al (2009) Metabolic analysis of the soil microbe *Dechloromonas aromatica* str. RCB: indications of a surprisingly complex life-style and cryptic anaerobic pathways for aromatic degradation. BMC Genomics 10:351
363. Cravo-Laureau C, Matheron R, Cayol J-L, Joulian C, Hirschler-Rea A (2004) *Desulfatibacillum aliphaticivorans* gen. nov., sp. nov., an *n*-alkane- and *n*-alkene-degrading, sulfate-reducing bacterium. Int J Syst Evol Microbiol 54:77–83
364. Cravo-Laureau C, Matheron R, Joulian C, Cayol J-L, Hirschler-Réa A (2004) *Desulfatibacillum alkenivorans* sp. nov., a novel n-alkene-degrading, sulfate-reducing bacterium, and emended description of the genus *Desulfatibacillum*. Int J Syst Evol Microbiol 54:1639–1642
365. Cravo-Laureau C, Labat C, Joulian C, Matheron R, Hirschler-Rea A (2007) *Desulfatiferula olefinivorans* gen. nov., sp. nov., a long-chain *n*-alkene-degrading, sulfate-reducing bacterium. Int J Syst Evol Microbiol 57:2699–2702
366. Kunapuli U, Jahn MK, Lueders T, Geyer R, Heipieper HJ, Meckenstock RU (2010) *Desulfitobacterium aromaticivorans* sp. nov. and *Geobacter toluenoxydans* sp. nov., iron-reducing bacteria capable of anaerobic degradation of monoaromatic hydrocarbons. Int J Syst Evol Microbiol 60:686–695
367. Rabus R, Nordhaus R, Ludwig W, Widdel F (1993) Complete oxidation of toluene under strictly anoxic conditions by a new sulfate-reducing bacterium. Appl Environ Microbiol 59:1444–1451
368. Kuever J, Rainey FA, Widdel F (2005) Genus III. *Desulfothermus* gen. nov. In: Brenner DJ et al (eds) Bergey's Manual of systematic bacteriology. Springer, New York
369. Rueter P, Rabus R, Wilkest H, Aeckersberg F et al (1994) Anaerobic oxidation of hydrocarbons in crude oil by new types of sulphate-reducing bacteria. Nature 372:455–458
370. Ommedal H, Torsvik T (2007) *Desulfotignum toluenicum* sp. nov., a novel toluene-degrading, sulphate-reducing bacterium isolated from an oil-reservoir model column. Int J Syst Evol Microbiol 57:2865–2869
371. Morasch B, Schink B, Tebbe C, Meckenstock R (2004) Degradation of *o*-xylene and *m*-xylene by a novel sulfate-reducer belonging to the genus *Desulfotomaculum*. Arch Microbiol 181:407–417
372. Yumoto I, Nakamura A, Iwata H, Kojima K et al (2002) *Dietzia psychralcaliphila* sp. nov., a novel, facultatively psychrophilic alkaliphile that grows on hydrocarbons. Int J Syst Evol Microbiol 52:85–90
373. Yousaf S, Afzal M, Reichenauer TG, Brady CL, Sessitsch A (2011) Hydrocarbon degradation, plant colonization and gene expression of alkane degradation genes by endophytic *Enterobacter ludwigii* strains. Environ Pollut 159:2675–2683
374. Mohanty G, Mukherji S (2008) Biodegradation rate of diesel range *n*-alkanes by bacterial cultures *Exiguobacterium aurantiacum* and *Burkholderia cepacia*. Int Biodeterior Biodegrad 61:240–250
375. Šepič E, Bricelj M, Leskovšek H (1997) Biodegradation studies of polyaromatic hydrocarbons in aqueous media. J Appl Microbiol 83:561–568
376. Rodríguez-Blanco A, Vetion G, Escande M-L, Delille D, Ghiglione J-F (2010) *Gallaecimonas pentaromativorans* gen. nov., sp. nov., a bacterium carrying 16S rRNA gene heterogeneity and able to degrade high-molecular-mass polycyclic aromatic hydrocarbons. Int J Syst Evol Microbiol 60:504–509
377. Lavania M, Cheema S, Sarma P, Mandal A, Lal B (2012) Biodegradation of asphalt by *Garciaella petrolearia* TERIG02 for viscosity reduction of heavy oil. Biodegradation 23:15–24
378. Coates JD, Bhupathiraju VK, Achenbach LA, McInerney MJ, Lovley DR (2001) *Geobacter hydrogenophilus*, *Geobacter chapellei* and *Geobacter grbiciae*, three new, strictly anaerobic, dissimilatory Fe(III)-reducers. Int J Syst Evol Microbiol 51:581–588
379. Lovley DR, Giovannoni SJ, White DC, Champine JE et al (1993) *Geobacter metallireducens* gen. nov. sp. nov., a microorganism capable of coupling the complete oxidation of organic compounds to the reduction of iron and other metals. Arch Microbiol 159:336–344

380. Weelink SAB, Van Doesburg W, Saia FT, Rijpstra WIC et al (2009) A strictly anaerobic betaproteobacterium *Georgfuchsia toluolica* gen. nov., sp. nov. degrades aromatic compounds with Fe(III), Mn(IV) or nitrate as an electron acceptor. FEMS Microbiol Ecol 70:575–585
381. Kummer C, Schumann P, Stackebrandt E (1999) *Gordonia alkanivorans* sp. nov., isolated from tar-contaminated soil. Int J Syst Bacteriol 49:1513–1522
382. Mikolasch A, Hammer E, Schauer F (2003) Synthesis of imidazol-2-yl amino acids by using cells from alkane-oxidizing bacteria. Appl Environ Microbiol 69:1670–1679
383. Kästner M, Breuer-Jammali M, Mahro B (1994) Enumeration and characterization of the soil microflora from hydrocarbon-contaminated soil sites able to mineralize polycyclic aromatic hydrocarbons (PAH). Appl Microbiol Biotechnol 41:267–273
384. Bonfá MRL, Grossman MJ, Mellado E, Durrant LR (2011) Biodegradation of aromatic hydrocarbons by Haloarchaea and their use for the reduction of the chemical oxygen demand of hypersaline petroleum produced water. Chemosphere 84:1671–1676
385. Wang Q, Fang X, Bai B, Liang X et al (2007) Engineering bacteria for production of rhamnolipid as an agent for enhanced oil recovery. Biotechnol Bioeng 98:842–853
386. Sarma PM, Bhattacharya D, Krishnan S, Lal B (2004) Degradation of polycyclic aromatic hydrocarbons by a newly discovered enteric bacterium, *Leclercia adecarboxylata*. Appl Environ Microbiol 70:3163–3166
387. Gauthier MJ, Lafay B, Christen R, Fernandez L et al (1992) *Marinobacter hydrocarbonoclasticus* gen. nov., sp. nov., a new, extremely halotolerant, hydrocarbon-degrading marine bacterium. Int J Syst Bacteriol 42:568–576
388. Schippers A, Bosecker K, Spröer C, Schumann P (2005) *Microbacterium oleivorans* sp. nov. and *Microbacterium hydrocarbonoxydans* sp. nov., novel crude-oil-degrading Gram-positive bacteria. Int J Syst Evol Microbiol 55:655–660
389. Adebusoye SA, Amund OO, Ilori MO, Domeih DO, Okpuzor J (2008) Growth and biosurfactant synthesis by Nigerian hydrocarbon-degrading estuarine bacteria. Rev Biol Trop 56:1603–1611
390. Hennessee CT, Seo J-S, Alvarez AM, Li QX (2009) Polycyclic aromatic hydrocarbon-degrading species isolated from Hawaiian soils: *Mycobacterium crocinum* sp. nov., *Mycobacterium pallens* sp. nov., *Mycobacterium rutilum* sp. nov., *Mycobacterium rufum* sp. nov. and *Mycobacterium aromaticivorans* sp. nov. Int J Syst Evol Microbiol 59:378–387
391. Zobell CE (1946) Action of microörganisms on hydrocarbons. Microbiol Rev 10:1–49
392. Willumsen P, Karlson U, Stackebrandt E, Kroppenstedt RM (2001) *Mycobacterium frederiksbergense* sp. nov., a novel polycyclic aromatic hydrocarbon-degrading Mycobacterium species. Int J Syst Evol Microbiol 51:1715–1722
393. Bushnell LD, Haas HF (1941) The utilization of certain hydrocarbons by microörganisms. J Bacteriol 41:653–673
394. Derz K, Klinner U, Schuphan I, Stackebrandt E, Kroppenstedt RM (2004) *Mycobacterium pyrenivorans* sp. nov., a novel polycyclic-aromatic-hydrocarbon-degrading species. Int J Syst Evol Microbiol 54:2313–2317
395. Hedlund BP, Geiselbrecht AD, Bair TJ, Staley JT (1999) Polycyclic aromatic hydrocarbon degradation by a new marine bacterium, *Neptunomonas naphthovorans* gen. nov., sp. nov. Appl Environ Microbiol 65:251–259
396. Jamison VW, Raymond RL, Hudson JO (1969) Microbial hydrocarbon co-oxidation. III. Isolation and characterization of an α, α'-dimethyl-*cis*, *cis*-muconic acid-producing strain of *Nocardia corallina*. Appl Microbiol 17:853–856
397. Raymond RL, Jamison VW, Hudson JO (1967) Microbial hydrocarbon co-oxidation. I. Oxidation of mono- and dicyclic hydrocarbons by soil isolates of the genus *Nocardia*. Appl Microbiol 15:857–865
398. Kim J-S, Powalla M, Lang S, Wagner F, Lünsdorf H, Wray V (1990) Microbial glycolipid production under nitrogen limitation and resting cell conditions. J Biotechnol 13:257–266

399. Kubota M, Kawahara K, Sekiya K, Uchida T et al (2005) *Nocardioides aromaticivorans* sp. nov., a dibenzofuran-degrading bacterium isolated from dioxin-polluted environments. Syst Appl Microbiol 28:165–174
400. Balkwill DL, Drake GR, Reeves RH, Fredrickson JK et al (1997) Taxonomic study of aromatic-degrading bacteria from deep-terrestrial-subsurface sediments and description of *Sphingomonas aromaticivorans* sp. nov., *Sphingomonas subterranea* sp. nov., and *Sphingomonas stygia* sp. nov. Int J Syst Bacteriol 47:191–201
401. Romine MF, Stillwell LC, Wong K–K, Thurston SJ et al (1999) Complete sequence of a 184-kilobase catabolic plasmid from *Sphingomonas aromaticivorans* F199. J Bacteriol 181:1585–1602
402. Takeuchi M, Hamana K, Hiraishi A (2001) Proposal of the genus *Sphingomonas* sensu stricto and three new genera, *Sphingobium*, *Novosphingobium* and *Sphingopyxis*, on the basis of phylogenetic and chemotaxonomic analyses. Int J Syst Evol Microbiol 51:1405–1417
403. Yuan J, Lai Q, Zheng T, Shao Z (2009) *Novosphingobium indicum* sp. nov., a polycyclic aromatic hydrocarbon-degrading bacterium isolated from a deep-sea environment. Int J Syst Evol Microbiol 59:2084–2088
404. Suzuki S, Hiraishi A (2007) *Novosphingobium naphthalenivorans* sp. nov., a naphthalene-degrading bacterium isolated from polychlorinated-dioxin-contaminated environments. J Gen Appl Microbiol 53:221–228
405. Sohn JH, Kwon KK, Kang J-H, Jung H-B, Kim S-J (2004) *Novosphingobium pentaromativorans* sp. nov., a high-molecular-mass polycyclic aromatic hydrocarbon-degrading bacterium isolated from estuarine sediment. Int J Syst Evol Microbiol 54:1483–1487
406. Teramoto M, Ohuchi M, Hatmanti A, Darmayati Y et al (2011) *Oleibacter marinus* gen. nov., sp. nov., a bacterium that degrades petroleum aliphatic hydrocarbons in a tropical marine environment. Int J Syst Evol Microbiol 61:375–380
407. Yakimov MM, Giuliano L, Gentile G, Crisafi E et al (2003) *Oleispira antarctica* gen. nov., sp. nov., a novel hydrocarbonoclastic marine bacterium isolated from Antarctic coastal sea water. Int J Syst Evol Microbiol 53:779–785
408. Szabó I, Szoboszlay S, Kriszt B, Háhn J et al (2011) *Olivibacter oleidegradans* sp. nov., a hydrocarbon-degrading bacterium isolated from a biofilter clean-up facility on a hydrocarbon-contaminated site. Int J Syst Evol Microbiol 61:2861–2865
409. Engelhardt MA, Daly K, Swannell RPJ, Head IM (2001) Isolation and characterization of a novel hydrocarbon-degrading, Gram-positive bacterium, isolated from intertidal beach sediment, and description of *Planococcus alkanoclasticus* sp. nov. J Appl Microbiol 90:237–247
410. Das R, Tiwary BN (2013) Isolation of a novel strain of *Planomicrobium chinense* from diesel contaminated soil of tropical environment. J Basic Microbiol 53:723–732
411. Jeon CO, Park W, Ghiorse WC, Madsen EL (2004) *Polaromonas naphthalenivorans* sp. nov., a naphthalene-degrading bacterium from naphthalene-contaminated sediment. Int J Syst Evol Microbiol 54:93–97
412. Gutierrez T, Green DH, Nichols PD, Whitman WB, Semple KT, Aitken MD (2013) *Polycyclovorans algicola* gen. nov., sp. nov., an aromatic-hydrocarbon-degrading marine bacterium found associated with laboratory cultures of marine phytoplankton. Appl Environ Microbiol 79:205–214
413. Gutierrez T, Nichols PD, Whitman WB, Aitken MD (2012) *Porticoccus hydrocarbonoclasticus* sp. nov., an aromatic hydrocarbon-degrading bacterium identified in laboratory cultures of marine phytoplankton. Appl Environ Microbiol 78:628–637
414. Hedlund BP, Staley JT (2006) Isolation and characterization of *Pseudoalteromonas* strains with divergent polycyclic aromatic hydrocarbon catabolic properties. Environ Microbiol 8:178–182
415. Stanier RY, Palleroni NJ, Doudoroff M (1966) The aerobic pseudomonads a taxonomic study. J Gen Microbiol 43:159–271

416. Zylstra GJ, Gibson DT (1989) Toluene degradation by *Pseudomonas putida* F1. Nucleotide sequence of the todC1C2BADE genes and their expression in *Escherichia coli*. J Biol Chem 264:14940–14946
417. Kotturi G, Robinson CW, Inniss WE (1991) Phenol degradation by a psychrotrophic strain of *Pseudomonas putida*. Appl Microbiol Biotechnol 34:539–543
418. Juang R-S, Wu C-Y (2007) Microbial degradation of phenol in high-salinity solutions in suspensions and hollow fiber membrane contactors. Chemosphere 66:191–198
419. Ramsay B, McCarthy J, Guerra-Santos L, Kappeli O, Fiechter A, Margaritis A (1988) Biosurfactant production and diauxic growth of *Rhodococcus aurantiacus* when using *n*-alkanes as the carbon source. Can J Microbiol 34:1209–1212
420. Iwabuchi N, Sunairi M, Anzai H, Nakajima M, Harayama S (2000) Relationships between colony morphotypes and oil tolerance in *Rhodococcus rhodochrous*. Appl Environ Microbiol 66:5073–5077
421. Austin B, Calomiris JJ, Walker JD, Colwell RR (1977) Numerical taxonomy and ecology of petroleum-degrading bacteria. Appl Environ Microbiol 34:60–68
422. Walker J, Colwell R (1974) Microbial degradation of model petroleum at low temperatures. Microb Ecol 1:63–95
423. Rentz JA, Alvarez PJJ, Schnoor JL (2008) Benzo[a]pyrene degradation by *Sphingomonas yanoikuyae* JAR02. Environ Pollut 151:669–677
424. Mallick S, Dutta TK (2008) Kinetics of phenanthrene degradation by *Staphylococcus* sp. strain PN/Y involving 2-hydroxy-1-naphthoic acid in a novel metabolic pathway. Process Biochem 43:1004–1008
425. Juhasz AL, Stanley GA, Britz ML (2000) Microbial degradation and detoxification of high molecular weight polycyclic aromatic hydrocarbons by *Stenotrophomonas maltophilia* strain VUN 10,003. Lett Appl Microbiol 30:396–401
426. Balachandran C, Duraipandiyan V, Balakrishna K, Ignacimuthu S (2012) Petroleum and polycyclic aromatic hydrocarbons (PAHs) degradation and naphthalene metabolism in *Streptomyces* sp. (ERI-CPDA-1) isolated from oil contaminated soil. Bioresour Technol 112:83–90
427. Kodama Y, Stiknowati LI, Ueki A, Ueki K, Watanabe K (2008) *Thalassospira tepidiphila* sp. nov., a polycyclic aromatic hydrocarbon-degrading bacterium isolated from seawater. Int J Syst Evol Microbiol 58:711–715
428. Zhao B, Wang H, Li R, Mao X (2010) *Thalassospira xianhensis* sp. nov., a polycyclic aromatic hydrocarbon-degrading marine bacterium. Int J Syst Evol Microbiol 60:1125–1129
429. Evans PJ, Mang DT, Kim KS, Young LY (1991) Anaerobic degradation of toluene by a denitrifying bacterium. Appl Environ Microbiol 57:1139–1145
430. Song B, Young LY, Palleroni NJ (1998) Identification of denitrifier strain T1 as *Thauera aromatica* and proposal for emendation of the genus *Thauera* definition. Int J Syst Bacteriol 48:889–894
431. Shinoda Y, Sakai Y, Uenishi H, Uchihashi Y et al (2004) Aerobic and anaerobic toluene degradation by a newly isolated denitrifying bacterium, *Thauera* sp. strain DNT-1. Appl Environ Microbiol 70:1385–1392
432. Perry J (2006) The genus *Thermoleophilum*. In: Dworkin M et al (eds) The prokaryotes. Springer, New York
433. Zarilla KA, Perry JJ (1984) *Thermoleophilum album* gen. nov. and sp. nov., a bacterium obligate for thermophily and *n*-alkane substrates. Arch Microbiol 137:286–290
434. Perry J (2006) The genus *Thermomicrobium*. In: Dworkin M et al (eds) The prokaryotes. Springer, New York
435. Feitkenhauer H, Müller R, Märkl H (2003) Degradation of polycyclic aromatic hydrocarbons and long chain alkanes at 60–70°C by *Thermus* and *Bacillus* spp. Biodegradation 14:367–372

436. Harwati TU, Kasai Y, Kodama Y, Susilaningsih D, Watanabe K (2008) *Tranquillimonas alkanivorans* gen. nov., sp. nov., an alkane-degrading bacterium isolated from Semarang Port in Indonesia. Int J Syst Evol Microbiol 58:2118–2121
437. Harwati TU, Kasai Y, Kodama Y, Susilaningsih D, Watanabe K (2009) *Tropicimonas isoalkanivorans* gen. nov., sp. nov., a branched-alkane-degrading bacterium isolated from Semarang Port in Indonesia. Int J Syst Evol Microbiol 59:388–391
438. Harwati TU, Kasai Y, Kodama Y, Susilaningsih D, Watanabe K (2009) *Tropicibacter naphthalenivorans* gen. nov., sp. nov., a polycyclic aromatic hydrocarbon-degrading bacterium isolated from Semarang Port in Indonesia. Int J Syst Evol Microbiol 59:392–396
439. Hedlund BP, Staley JT (2001) *Vibrio cyclotrophicus* sp. nov., a polycyclic aromatic hydrocarbon (PAH)-degrading marine bacterium. Int J Syst Evol Microbiol 51:61–66
440. Geiselbrecht AD, Herwig RP, Deming JW, Staley JT (1996) Enumeration and phylogenetic analysis of polycyclic aromatic hydrocarbon-degrading marine bacteria from Puget sound sediments. Appl Environ Microbiol 62:3344–3349
441. Lang S, Gilbon A, Syldatk C, Wagner F (1984) Comparison of interfacial active properties of glycolipis from microorganisms. In: Mittal KL, Lindman B (eds) Surfactants in solution. Plenum Press, New York
442. Sar N, Rosenberg E (1983) Emulsifier production by *Acinetobacter calcoaceticus* strains. Curr Microbiol 9:309–313
443. Barkay T, Navon-Venezia S, Ron EZ, Rosenberg E (1999) Enhancement of solubilization and biodegradation of polyaromatic hydrocarbons by the bioemulsifier alasan. Appl Environ Microbiol 65:2697–2702
444. Navon-Venezia S, Zosim Z, Gottlieb A, Legmann R et al (1995) Alasan, a new bioemulsifier from *Acinetobacter radioresistens*. Appl Environ Microbiol 61:3240–3244
445. Käppeli O, Finnerty WR (1979) Partition of alkane by an extracellular vesicle derived from hexadecane-grown Acinetobacter. J Bacteriol 140:707–712
446. Salehizadeh H, Mohammadizad S (2009) Microbial enhanced oil recovery using biosurfactant produced by *Alcaligenes faecalis*. Iran J Biotechnol 7:216–223
447. Passeri A, Schmidt M, Haffner T, Wray V, Lang S, Wagner F (1992) Marine biosurfactants. IV. Production, characterization and biosynthesis of an anionic glucose lipid from the marine bacterial strain MM1. Appl Microbiol Biotechnol 37:281–286
448. Zheng C, Li Z, Su J, Zhang R, Liu C, Zhao M (2012) Characterization and emulsifying property of a novel bioemulsifier by *Aeribacillus pallidus* YM-1. J Appl Microbiol 113:44–51
449. Li Z-Y, Lang S, Wagner F, Witte L, Wray V (1984) Formation and identification of interfacial-active glycolipids from resting microbial cells. Appl Environ Microbiol 48:610–617
450. Rosenberg E, Zuckerberg A, Rubinovitz C, Gutnick DL (1979) Emulsifier of *Arthrobacter* RAG-1: isolation and emulsifying properties. Appl Environ Microbiol 37:402–408
451. Cooper DG, Goldenberg BG (1987) Surface-active agents from two *Bacillus* species. Appl Environ Microbiol 53:224–229
452. Bordoloi NK, Konwar BK (2008) Microbial surfactant-enhanced mineral oil recovery under laboratory conditions. Colloids Surf B Biointerfaces 63:73–82
453. Horowitz S, Currie JK (1990) Novel dispersants of silicon carbide and aluminum nitride. J Dispers Sci Technol 11:637–659
454. McInerney MJ, Jenneman GE, Knapp RM, Menzie DE (1985) Biosurfactant and enhanced oil recovery. United States, US Patent 4522261
455. Folmsbee M, Duncan K, Han SO, Nagle D, Jennings E, McInerney M (2006) Re-identification of the halotolerant, biosurfactant-producing *Bacillus licheniformis* strain JF-2 as *Bacillus mojavensis* strain JF-2. Syst Appl Microbiol 29:645–649
456. Cooper DG, Zajic JE (1980) Surface-active compounds from microorganisms. In: Perlman D (ed) Advances in applied microbiology. Academic Press, New York
457. Gurjar M, Khire JM, Khan MI (1995) Bioemulsifier production by *Bacillus stearothermophilus* VR-8 isolate. Lett Appl Microbiol 21:83–86

458. Arima K, Kakinuma A, Tamura G (1968) Surfactin, a crystalline peptidelipid surfactant produced by *Bacillus subtilis*: Isolation, characterization and its inhibition of fibrin clot formation. Biochem Biophys Res Commun 31:488–494
459. Nitschke M, Pastore GM (2006) Production and properties of a surfactant obtained from *Bacillus subtilis* grown on cassava wastewater. Bioresour Technol 97:336–341
460. Oberbremer A, Müller-Hurtig R (1989) Aerobic stepwise hydrocarbon degradation and formation of biosurfactants by an original soil population in a stirred reactor. Appl Microbiol Biotechnol 31:582–586
461. Oberbremer A, Müller-Hurtig R, Wagner F (1990) Effect of the addition of microbial surfactants on hydrocarbon degradation in a soil population in a stirred reactor. Appl Microbiol Biotechnol 32:485–489
462. McCaffrey WC, Cooper DG (1995) Sophorolipids production by *Candida bombicola* using self-cycling fermentation. J Ferment Bioeng 79:146–151
463. Cirigliano MC, Carman GM (1984) Isolation of a bioemulsifier from *Candida lipolytica*. Appl Environ Microbiol 48:747–750
464. Cooper DG, Zajic JE, Gerson DF (1979) Production of surface-active lipids by *Corynebacterium lepus*. Appl Environ Microbiol 37:4–10
465. Kebbouche-Gana S, Gana M, Khemili S, Fazouane-Naimi F et al (2009) Isolation and characterization of halophilic *Archaea* able to produce biosurfactants. J Ind Microbiol Biotechnol 36:727–738
466. Calvo C, Martinez-Checa F, Mota A, Bejar V, Quesada E (1998) Effect of cations, pH and sulfate content on the viscosity and emulsifying activity of the *Halomonas eurihalina* exopolysaccharide. J Ind Microbiol Biotechnol 20:205–209
467. Trebbau de Acevedo G, McInerney MJ (1996) Emulsifying activity in thermophilic and extremely thermophilic microorganisms. J Ind Microbiol 16:1–7
468. Zhang Y, Miller RM (1992) Enhanced octadecane dispersion and biodegradation by a *Pseudomonas* rhamnolipid surfactant (biosurfactant). Appl Environ Microbiol 58:3276–3282
469. Parra J, Guinea J, Manresa M, Robert M et al (1989) Chemical characterization and physicochemical behavior of biosurfactants. J Am Oil Chem Soc 66:141–145
470. Tahzibi A, Kamal F, Assadi MM (2004) Improved production of rhamnolipids by a *Pseudomonas aeruginosa* mutant. Iran Biomed J 8:25–31
471. Raza Z, Rehman A, Khan M, Khalid Z (2007) Improved production of biosurfactant by a *Pseudomonas aeruginosa* mutant using vegetable oil refinery wastes. Biodegradation 18:115–121
472. Ishigami Y, Gama Y, Ishii F, Choi YK (1993) Colloid chemical effect of polar head moieties of a rhamnolipid-type biosurfactant. Langmuir 9:1634–1636
473. Tuleva BK, Ivanov GR, Christova NE (2002) Biosurfactant production by a new *Pseudomonas putida* strain. Zeitschrift für Naturforschung, 57c:356–360
474. Mercade M, Manresa M (1994) The use of agroindustrial by-products for biosurfactant production. J Am Oil Chem Soc 71:61–64
475. Syldatk C, Lang S, Matulovic U, Wagner F (1985) Production of four interracial active rhamnolipids from *n*-alkanes or glycerol by resting cells of *Pseudomonas species* DSM 2874. Zeitschrift für Naturforschung, 40c:61–67
476. Ciapina E, Melo W, Santa AL, Santos A, Freire D, Pereira N (2006) Biosurfactant production by *Rhodococcus erythropolis* grown on glycerol as sole carbon source. Appl Biochem Biotechnol 131:880–886
477. Kretschmer A, Bock H, Wagner F (1982) Chemical and physical characterization of interfacial-active lipids from *Rhodococcus erythropolis* grown on *n*-alkanes. Appl Environ Microbiol 44:864–870
478. Espuny M, Egido S, Rodón I, Manresa A, Mercadé M (1996) Nutritional requirements of a biosurfactant producing strain *Rhodococcus* sp 51T7. Biotechnol Lett 18:521–526

479. Ramsay BA, Margaritias A, Zajic JE, Cooper DG (1983) *Rhodochrous* bacteria: biosurfactant production and demulsifying ability. In: Zajic JE et al (eds) Microbial enhanced oil recovery. Pennwell Publishing Co, Tulsa
480. Oh SG, Shah DO (1993) Effect of counterions on the interfacial tension and emulsion droplet size in the oil/water/dodecyl sulfate system. J Phys Chem 97:284–286
481. Hommel R, Stiiwer O, Stuber W, Haferburg D, Kleber HP (1987) Production of water-soluble surface-active exolipids by *Torulopsis apicola*. Appl Microbiol Biotechnol 26:199–205
482. Hommel RK (1990) Formation and physiological role of biosurfactants produced by hydrocarbon-utilizing microorganisms. Biodegradation 1:107–119
483. Stüwer O, Hommel R, Haferburg D, Kleber HP (1987) Production of crystalline surface-active glycolipids by a strain of *Torulopsis apicola*. J Biotechnol 6:259–269
484. Cooper DG, Paddock DA (1984) Production of a biosurfactant from *Torulopsis bombicola*. Appl Environ Microbiol 47:173–176
485. Göbbert U, Lang S, Wagner F (1984) Sophorose lipid formation by resting cells of *Torulopsis bombicola*. Biotechnol Lett 6:225–230

Adv Biochem Eng Biotechnol (2014) 142: 95–121
DOI: 10.1007/10_2013_257

Published Online: 6 December 2013

Relevance of Deep-Subsurface Microbiology for Underground Gas Storage and Geothermal Energy Production

Claudia Gniese, Petra Bombach, Jana Rakoczy, Nils Hoth, Michael Schlömann, Hans-Hermann Richnow and Martin Krüger

Abstract This chapter gives the reader an introduction into the microbiology of deep geological systems with a special focus on potential geobiotechnological applications and respective risk assessments. It has been known for decades that microbial activity is responsible for the degradation or conversion of hydrocarbons in oil, gas, and coal reservoirs. These processes occur in the absence of oxygen, a typical characteristic of such deep ecosystems. The understanding of the responsible microbial processes and their environmental regulation is not only of great scientific interest. It also has substantial economic and social relevance, inasmuch as these processes directly or indirectly affect the quantity and quality of the stored

C. Gniese · M. Schlömann
TU Bergakademie Freiberg, Institute of Biosciences, Leipziger Str. 29, 09599 Freiberg, Germany
e-mail: claudia.gniese@ioez.tu-freiberg.de

M. Schlömann
e-mail: michael.schloemann@ioez.tu-freiberg.de

P. Bombach · H.-H. Richnow
Department of Isotope Biogeochemistry, Helmholtz Centre for Environmental Research-UFZ, Permoserstr. 15, 04318 Leipzig, Germany
e-mail: petra.bombach@ufz.de

H.-H. Richnow
e-mail: hans.richnow@ufz.de

J. Rakoczy · M. Krüger (✉)
Federal Institute for Geosciences and Natural Resources-BGR, Stilleweg 2, 30655 Hannover, Germany
e-mail: martin.krueger@bgr.de

N. Hoth
TU Bergakademie Freiberg, Institute of Mining, Gustav-Zeuner-Str. 1A, 09599 Freiberg, Germany
e-mail: nils.hoth@tu-freiberg.de

oil or gas. As outlined in the following chapter, in addition to the conventional hydrocarbons, new interest in such deep subsurface systems is rising for different technological developments. These are introduced together with related geomicrobiological topics. The capture and long-term storage of large amounts of carbon dioxide, carbon capture and storage (CCS), for example, in depleted oil and gas reservoirs, is considered to be an important option to mitigate greenhouse gas emissions and global warming. On the other hand, the increasing contribution of energy from natural and renewable sources, such as wind, solar, geothermal energy, or biogas production leads to an increasing interest in underground storage of renewable energies. Energy carriers, that is, biogas, methane, or hydrogen, are often produced in a nonconstant manner and renewable energy may be produced at some distance from the place where it is needed. Therefore, storing the energy after its conversion to methane or hydrogen in porous reservoirs or salt caverns is extensively discussed. All these developments create new research fields and challenges for microbiologists and geobiotechnologists. As a basis for respective future work, we introduce the three major topics, that is, CCS, underground storage of gases from renewable energy production, and the production of geothermal energy, and summarize the current state of knowledge about related geomicrobiological and geobiotechnological aspects in this chapter. Finally, recommendations are made for future research.

Keywords CCS · Deep biosphere · Geothermal energy · Hydrocarbon reservoir · Renewable energy · Underground gas storage

Abbreviations

16S rRNA	Ribosomal RNA of a sedimentation rate of 16 Svedberg
AOM	Anaerobic oxidation of methane
bbl	Barrel (oil)
CARD-FISH	Catalyzed reporter deposition-Fluorescence in situ hybridisation
CCS	Carbon capture and storage
cDNA	Complementary DNA
CLEAN	CO_2 large-scale enhanced gas recovery in the Altmark Natural Gas Field
CO_2CRC	Cooperative Research Centre for Greenhouse Gas Technologies
COE	Cost of electricity
CO_2-EGR	EGR using CO_2
CO_2-EOR	EOR using CO_2
CO_2MAN	CO_2-reservoir management
CO_2SINK	CO_2 Storage by injection into a saline aquifer at Ketzin
DAPI	4′,6-Diamidin-2-phenylindol
DGGE	Denaturing gradient gel electrophoresis
DNA	Deoxyribonucleic acid
DOC	Dissolved organic carbon
*dsr*AB	Dissimilatory (bi)sulfite reductase gene

EDTA	Ethylenediaminetetraacetic acid
ECBM	Enhanced coal bed mining
EGR	Enhanced gas recovery
EOR	Enhanced oil recovery
EPS	Extracellular polymeric substances
FISH	Fluorescence in situ hybridization
HFC	Hydrofluorocarbons
IEAGHG	International Energy Agency Greenhouse Gas
mcr	Methyl coenzyme M reductase gene
MEOR	Microbial enhanced oil recovery
MIC	Microbially influenced corrosion
MPN	Most probable number
mRNA	Messenger RNA
OIP	Oil in place
P	Pressure
PAH	Polycyclic aromatic hydrocarbons
PFC	Perfluorocarbons
PCR	Polymerase chain reaction
PDS	Bottom-hole positive displacement sampler
PLFA	Phospholipid-derived fatty acids
qPCR	Quantitative polymerase chain reaction
RECOBIO	Recycling of sequestrated CO_2 by deep subsurface microbial-biogeochemical transformation, RECOBIO-1 and RECOBIO-2 are two successive projects
RNA	Ribonucleic acid
RT-qPCR	Reverse transcription-quantitative PCR
SAC	Surface active compound
SC-CO_2	Supercritical carbon dioxide
SIP	Stable isotope probing
SSCP	Single-strand conformation polymorphism
TOC	Total organic carbon
T-RFLP	Terminal restriction fragment length polymorphism
UGS	Underground gas storage
V	Volume

Contents

1 Carbon Capture and Storage (CCS) 98
1.1 Introduction to Carbon Capture and Storage 98
1.2 Geochemical Effects and Risks of CO_2 in Storage Sites 100
1.3 Microbial Populations in Potential CO_2 Storage Sites 102
1.4 Conclusion and Perspectives 107

2 Underground Gas Storage (Methane, Hydrogen) for Energy Generation 108
2.1 Introduction to Underground Gas Storage 108
2.2 Microbiology of Gas Storage Sites 109
2.3 Implications and Future Perspectives 112
3 Geothermal Energy Production 112
3.1 Geothermal Energy 112
3.2 Geothermal Energy and its Effects on Subsurface Microbiology 113
3.3 Further Research 116
References 116

1 Carbon Capture and Storage (CCS)

1.1 Introduction to Carbon Capture and Storage

Mitigation of greenhouse gas emissions without interference with economic growth is the main concern of climate-change initiatives to prevent global warming. Unfortunately, observations of a 100-year period between 1906 and 2005 already show an increase of the global temperature of 0.74 ± 0.18 °C. Changes in climate are noticeable and include extreme weather such as droughts, heavy precipitation, heat waves, and intensity of tropical cyclones [45].

CO_2 is the principal component of the greenhouse gases in addition to CH_4, N_2O, hydrofluorocarbons (HFC), perfluorocarbons (PFC), and SF_6 (Kyoto Protocol, 1998). Power generation using fossil fuels or biomass, cement production, and other CO_2-emitting industries are the main sources of CO_2. This gas accounts for 64 % of the enhanced "greenhouse effect" [15, 44]. Therefore, removing CO_2 from flue gases would help to maintain the global temperature rise to a maximum of 2 °C.

In this respect, carbon capture and storage (CCS) can be a promising and fast approach to reduce CO_2 emission to the atmosphere. But this approach is limited by availability and capacity of CO_2 storage sites. Despite this limitation, CCS can be a bridging technology that provides a gain in time until an energy supply with renewable energies is secured. Moreover, storage of CO_2 in deep geological formations probably results in natural gas restoration in geological timescales provided that CO_2 is transformed microbiologically to CH_4.

In the special report of the Intergovernmental Panel on Climate Change on carbon dioxide capture and storage [44], CCS is defined as "[…] a process consisting of the separation of CO_2 from industrial and energy-related sources, transport to a storage location and long-term isolation from the atmosphere." A detailed description of the CCS technology is given in this special report. In brief, there are three main strategies to capture CO_2 from flue gases: (i) postcombustion, (ii) precombustion, and (iii) oxy–fuel combustion [44]. In the postcombustion process, chemical sorbents are used to recover up to 95 % CO_2 from the flue gas, which contains mainly N_2 and 3–15 vol % CO_2. In the precombustion process, the fuel is first burned with oxygen, air, and/or steam to generate CO and H_2. Then CO is converted to CO_2 by the addition of steam and finally CO_2 is captured using absorption–desorption methods. In the oxy–fuel combustion process, the combustion of the fuel is carried out by using oxygen, either pure or mixed with a CO_2-rich recycled flue gas, and results in flue gas of up to 98 % CO_2. After CO_2 is enriched from the original flue gas by using one of these capturing strategies, the gas is pressurized for the transport to CO_2 storage sites via pipelines or trucks. The CO_2 capture process accounts for an increase of 20–90 % cost of electricity (COE) depending on the type of power plant [44]. Further technological developments may reduce the extra costs in the future. The sequestration of the original flue gas would cause much higher costs.

There are research and industrial projects worldwide that investigate CCS on laboratory and field scales (pilot/demonstration plants) and perform EGR (enhanced gas recovery), EOR (enhanced oil recovery), or ECBM (enhanced coal bed mining) connected to CO_2 storage. The IEAGHG [43] operates a database that lists all research, development, and demonstration (RD&D) projects concerning CCS. Among them, there are projects that store CO_2 in saline aquifers, for example, the Frio Brine Pilot Test (USA) and CO_2SINK (Ketzin, Germany) projects, which store CO_2 in gas fields, for example, the In Salah Gas project (Algeria) and CO_2CRC Otway Basin project (Australia), and EOR projects, for example, Weyburn CO_2-EOR (Canada). In addition, a CO_2-EGR approach was planned for the gas field Altmark (CLEAN project, Germany). The almost depleted gas field Altmark is the second-largest onshore gas field in Europe and would be of great importance for CO_2 storage if CCS is accepted by the German government and society. In addition to research and development in CCS technology, industry and scientists have to include good public relations in their field of duty. In particular in populated regions where CCS in deep geological formations is possible, residents must regularly be informed about the process, the safety, and the risk management of the CO_2 storage site.

Disposal of CO_2 in the ocean and usage of CO_2 for chemical processes are also approaches to reduce emissions of CO_2. But the most promising approach is the injection of supercritical CO_2 (SC-CO_2) into deep geological formations, that is, depleted oil and gas reservoirs, saline aquifers, or into unminable coal beds. In general, geological formations have to fulfill two main requirements to be suitable for long-term CO_2 storage. First, the storage reservoir has to consist of a porous and permeable rock, often sandstone, into which the CO_2 can be injected. Second, there has to be an impermeable cap rock and a succession of further seals up to the surface (multibarrier system). Typical cap rocks and seals consist of mudstone, siltstone, or salts (e.g., anhydrite). In particular, natural gas reservoirs have been demonstrated to be gas tight at least concerning CH_4 for geological timescales. Therefore, CO_2 storage in depleted gas fields is favored. These storage sites can be operated up to a site-specific pressure level, which should remain below the initial pressure level of the reservoir.

The worldwide storage potential has been estimated to be at least 200 Gt CO_2, and might even reach 2,000 Gt CO_2 in sedimentary basins (e.g., oil and gas reservoirs) [44]. For Germany, a summary of the distribution and the storage potential of sedimentary basins has been provided by May et al. [63]. There are three main geological structures, which represent potential CO_2 storage reservoirs in Germany. These reservoirs comprise sandstones rich in (i) feldspar, carbonate, and clay; (ii) iron minerals; or (iii) organic material. The formation waters are often highly saline (up to 300 g/l) and consist of high ammonia content (up to 3,000 ppm). With depth, the brines are increasingly reductive, their content of dissolved metal ions (e.g., iron) increases, whereas the content of sulfate decreases. Aside from these chemical conditions, the deep biosphere, which is likely to be present in such geological formations, has to be adapted to high temperatures, high pressure, and a low supply of electron acceptors, electron donors, or other nutrients.

1.2 Geochemical Effects and Risks of CO_2 in Storage Sites

The CO_2 gas designated for storage can be accompanied by impurities such as SO_x, NO_x, CO, H_2S, NH_3, O_2, condensable water, and hydrocarbons [87]. Therefore, the potential impact of the impurities on the storage site and storage process has to be considered. According to Knauss et al. [51], co-contaminant H_2S showed only minor effects on water–rock interaction, but SO_2 leads to a drastic drop of pH, which will lower the formation of carbonates. However, sequestration of CO_2–SO_2 mixtures into storage sites that contain hematite (Fe_2O_3, red beds) has been reported to result in dissolution of hematite and the release of ferrous iron induced by SO_2 [77]. This iron release will promote the formation of siderite ($FeCO_3$), which can cause an increase in the storage capacity of the reservoir, but can also provoke a negative effect on the storage unit by lowering its permeability:

$$Fe_2O_3 + 2\,CO_2(g) + SO_2(g) + H_2O \rightarrow 2\,FeCO_3 + H_2SO_4 \qquad (1)$$

Once CO_2 is injected into the storage reservoir, the gas can be trapped by four mechanisms [38, 77, 78]:

(i) Hydrodynamic trapping: SC–CO_2 is trapped below a cap rock of a depleted gas or oil field. This can be connected to enhanced gas or oil recovery (EGR, EOR), respectively.
(ii) Residual trapping: CO_2 is trapped by capillary forces in the pores of the reservoir rocks.
(iii) Solubility trapping/solution trapping: CO_2 is dissolved in formation water as H_2CO_3, HCO_3^-, and other aqueous species.
(iv) Mineral trapping: CO_2 is trapped as carbonate mineral (calcite, magnesite, siderite, and dawsonite) in deep saline formations. In this respect, silicate minerals are essential because their alteration enhances these mineral trapping processes due to the supply of cations.

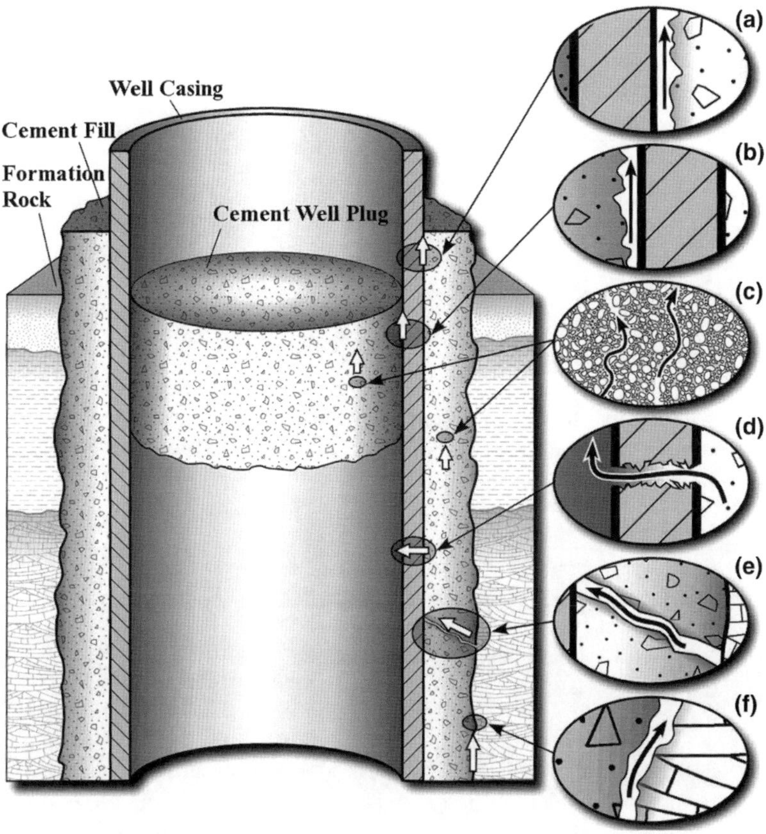

Fig. 1 Schematic illustration of possible leakage pathways through an abandoned well. **a** Between casing and cement; **b** between cement plug and casing; **c** through the cement porespace as a result of cement degradation; **d** through casing as a result of corrosion; **e** through fractures in cement; and **f** between cement and rock (from [33], with kind permission from Springer Science and Business Media: [33], Fig. 1, Copyright Springer-Verlag 2004)

Another trapping mechanism can be the absorption of CO_2 by coal, which could lead to a release of methane and is used in enhanced coal bed methane recovery.

It has to be considered that CO_2 differs from other gases with respect to its solubility, penetration, and reaction behavior. The permeability and penetration behavior of the cap rock is also a crucial aspect for the safety and integrity of the CO_2 storage site. One possible risk could be leakage of CO_2 via undetected fractures and faults and via abandoned wells or failure during the injection process [44]. However, CO_2 exhibits a very good solubility (in contrast to CH_4 and especially to H_2) and will be trapped in any overlying formation water if it leaks vertically through one sealing unit. Leakage through anthropogenic artificial barriers (cement, casing) may occur because of fatigue or alteration of the well bore material due to chemical attack of highly corrosive SC–CO_2 or high pressure (Fig. 1 [33]).

Another risk, especially for CO_2 storage in saline aquifers, could be contamination of overlying groundwater with brines. The saline formation water could be displaced upward due to a spacious pressure build-up. In this case, the pressure of the storage formation would drop, could be detected with monitoring equipment, and an emergency plan could be applied. In general, monitoring strategies have to be operated before, during, and after CO_2 sequestration to assess the baseline conditions, to follow the storage process and detect process failure, and to control long-term reactions and failure, respectively. The migration of CO_2 in the storage formation and the composition of the overlying groundwater and surface soils of the storage site have to be controlled. In addition to geochemical reactions, also biogeochemical reactions, that is, mineral–brine–CO_2–microbe interactions have to be considered.

1.3 Microbial Populations in Potential CO_2 Storage Sites

Geological formations known to be suitable for CCS comprise a deep subsurface biosphere, which is dominated by sulfate-reducing, iron-reducing, acetogenic, and methanogenic microorganisms [60]. Microbial corrosion of tubing and cement of well bores and souring of gas due to H_2S production by sulfate reducers are well-known problems of gas and oil industry [21, 34, 50]. In addition, clogging of well bores and porespace of the geological formation can arise when H_2S precipitates in the presence of ferrous iron first to FeS and then to FeS_2. These technological problems clarify the need to consider biogeochemical reactions in addition to geochemical reactions, although microbiologically mediated processes in the deep subsurface are rather slow compared to microbial activities at the surface [17, 60].

Microbial reactions can have favorable and unfavorable effects on the capacity, integrity, and safety of CO_2 storage sites. Therefore, baseline monitoring of each CCS operation should include the detection of the initial microbial community to deduce possible microbial reactions in advance. In particular, a microbial assemblage as biofilm on mineral surfaces can either inhibit or enhance mineral dissolution [64]. Dissolution of minerals can decrease the storage capacity of the reservoir and could additionally lead to exposure of fractures, which possibly form connections to higher layers of the formation and affect the storage integrity. But dissolution of minerals can also provide microorganisms with, for example, electron acceptors. Biofilms can serve as protective coating of minerals decreasing mineral dissolution and presenting nucleation sites to catalyze carbonate precipitation [23]. On the surface of silica-based minerals, for example, the formation of amorphous silica gels, whose crosslinking is facilitated in the presence of biomolecules, can lead to a self-sealing effect of microfissures and porespace of disturbed claystone and cements [36, 49]. Hence, self-sealing and enhanced carbonate formation may contribute significantly to integrity and safety of the storage site and additionally stabilize the injected CO_2 into solid carbonates [64]. Another indirect way to favor carbonation can be the adjustment of physicochemical conditions (e.g., increase of pH) due to metabolic activity in the deep subsurface.

The injection of CO_2 into potential storage formations causes changes in reservoir temperature and pressure, and also leads to considerably higher CO_2 concentrations. All variations in the physicochemical conditions will stress the indigenous biosphere of the storage formation. Beyond that, a sterilization can take place at the center of the CO_2 injection well. However, Mitchell et al. [66, 67] have demonstrated that the resilience of biofilms to SC–CO_2 is higher than that of planktonic microorganisms.

Microbial monitoring before and during CO_2 injection into a saline aquifer near Ketzin (Germany) has revealed that the microorganisms adapted within five months to higher CO_2 concentrations and were even more metabolically active [70]. Furthermore, during the propagation of CO_2 in the storage reservoir, CO_2 will form a plume that develops a gradient of CO_2 concentrations. Thus, regions with lower CO_2 content can directly provide autotrophic microorganisms with their carbon source and an electron acceptor. Heterotrophic microorganisms probably metabolize organic compounds (e.g., organic acids, methylalkanes) that were mobilized by SC–CO_2 from the sandstone of the storage formation [85]. Hence, CCS can even stimulate microbial growth.

The consumption of CO_2 due to microbial activity has reproducibly been shown to be connected to a considerable increase in the formation of TOC (total organic carbon) in experiments with a bioreactor and a sterile control reactor under elevated H_2 and CO_2 partial pressure [26]. The experiments have been performed with milled material of a drilling core, which originated from the gas field Schneeren-Husum (Germany), and formation water collected at well heads of well bores of this gasfield. Therefore, microbial transformation of CO_2 into biomass and organic compounds can additionally contribute to the storage capacity of a reservoir.

One problem that may result from the stimulation of, in particular, sulfate-reducing microorganisms is the increase in H_2S production, which in turn can affect the integrity of well bores and storage equipment via biocorrosion.

Methanogenic microorganisms form another microbial group to be considered, which would transform injected CO_2 either directly (autotrophically) or indirectly (acetoclastically) to CH_4. Although CH_4 represents a far more potent greenhouse gas than CO_2 if it would leak from the reservoir, CH_4 can possibly be used as an energy source in geological timescales.

So far, the extent of the microbial impact on CCS on short-term and long-term scales remains to be clarified. Even if no viable microorganism survives the CO_2 injection, there would be biological residues such as endospores, organic clusters, enzymes, or lysed cells that can have an influence on the CO_2 storage performance [62, 64]. There are many biogeochemical processes in the deep subsurface that are not yet understood or have even been subjected to investigation. In this respect, one challenge is to obtain reliable samples of the deep subsurface biosphere. Then, other challenging aspects are the very low doubling times of these microorganisms and the creation of their physicochemical requirements for cultivation. Despite these aspects, only a small number of CCS projects to date consider biogeochemical processes [17].

In some projects, which store CO_2 in hydrocarbon reservoirs, microbial monitoring of surface soil (total cell counts, In Salah Gas project; Algeria [47]) or microbial mats above the storage reservoir (microbial community compositon and total cell counts, Sleipner project (Norway [98]) has been performed to survey possible CO_2 leakage. In contrast, formation waters of the Paaratte formation have been sampled in situ at 1,400 m depth (60 °C, 13.8 MPa) for 16S rRNA gene analyses in the framework of the CO_2CRC Otway Basin project (Australia). Bacterial sequences of the reservoir community have been related to the genera *Thermincola*, *Acinetobacter*, *Sphingobium*, and *Dechloromonas* [72]. Microorganisms, stained with the DNA-specific dye DAPI, have been reported to be microscopically visible mainly as filamentous cells of 5–45 μm length. The injection of a gas mixture of 75.4 mol % CO_2 and 20.5 mol % CH_4 to the Paaratte formation started in 2008 [13].

Detailed microbial analyses have been performed for formation fluids of the almost depleted natural gas reservoir Altmark (Permian—Upper Rotliegend, Germany). This gas reservoir comprises extreme environmental conditions, for example, in situ temperatures of 110 °C up to 130 °C and high salinity brines of >300 g salts per liter.

The hydraulic isolated subfield block "Altensalzwedel" has been considered for EGR and storage of 100,000 t CO_2 in 3,000 m depth (Fig. 2). Although injection of CO_2 in the Altmark gasfield was not possible due to political obstacles and public opposition, a comprehensive reservoir monitoring, which includes 16S rRNA gene analyses and cell quantification of the deep subsurface biosphere, has been performed during the CLEAN project [56].

Formation fluids of three different well bores of the subfield block "Altensalzwedel" (S10, S13, S17) have been sampled in situ using a double-ball lining sampler. Analyses of bacterial 16S rRNA genes of these fluids have revealed that the microorganisms at the site are related to hydrogenotrophic bacteria of *Hydrogenophaga* sp., *Acidovorax* sp., *Ralstonia* sp., and *Pseudomonas* sp. and to representatives from saline, hot, anoxic, and deep environments [69]. In addition, relatives of *Diaphorobacter* sp., a thiosulfate-oxidizing bacterium, were present in the formation fluid of one well bore (S17), and an uncultured biocorrosive thermophilic bacterium has been detected in fluids of two well bores (S13, S17). The formation fluids of one well bore (S10) have also been sampled with a bottom-hole positive displacement sampler (PDS). This sampling device can be inserted sterile and closed into the well bore, can be opened in the depth to collect the formation water in situ, and closed again to be moved out. In contrast, the double-ball lining is a more open system for in situ sampling. However, the two different sampling procedures have principally revealed the same microbial community structure in the formation water of well bore S10. In addition to the microorganisms, which

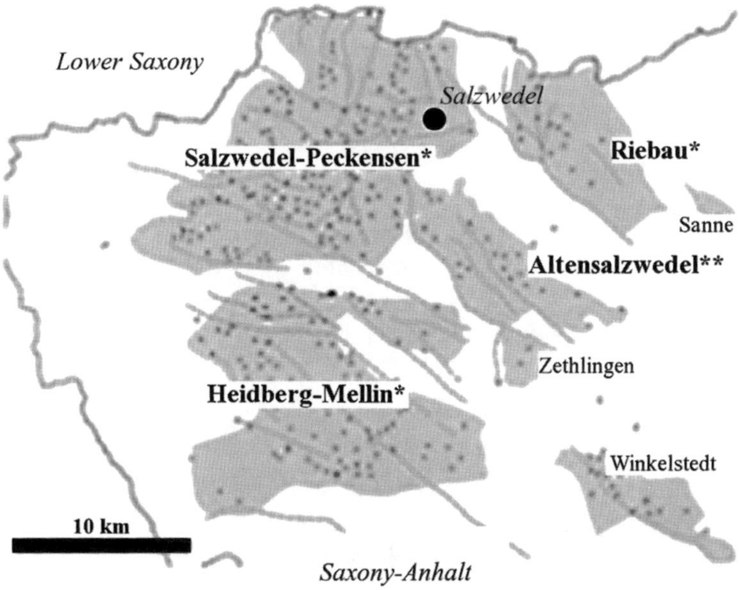

Fig. 2 Subfield blocks of the Altmark gasfield as indicated by *grey* shades (modified after [31]). *Three subfield blocks, which have been the focus of the RECOBIO-2 project. **One subfield block, which has been the focus of the CLEAN project

have also been found in the formation water sampled with the double-ball lining sampler, additional 16S rRNA gene sequences similar to dissimilatory metal-reducing bacteria (*Pantoea* sp. described by Francis et al. [32]), aromatic-degrading and metal-corroding bacteria of deep-sediment origin (*Sphingomonas* sp. [3, 99]), and extremophilic Fe(III)- and Mn(IV)-reducing bacteria (*Bacillus* sp. [12]) have been found in formation water sampled with the PDS [56]. Cell quantification using cell counting of SYBR Green-stained cells (mainly particle-associated cells) and quantitative PCR analyses have shown only very low cell numbers [69].

In the RECOBIO-2 project, the deep biosphere of three subfield blocks of the Altmark natural gasfield, surrounding the "Altensalzwedel" subfield block, has been investigated (Fig. 2) [42]. The formation waters of CLEAN and RECOBIO-2 sampling sites mainly differed in their concentration of sulfate, which was between 400–1,800 mg/l and almost no detectable sulfate, respectively.

Microscopic analyses using CARD-FISH and DAPI-stained cells also showed rather low cell numbers of at most 10^5 cells/ml in the formation water samples. There were only minor differences in the bacterial community composition in the formation water, which had been sampled at the well head (produced) and in situ (double-ball lining) of a well bore of the "Heidberg-Mellin" subfield block [35, 41]. The 16S rRNA gene sequences were similar to representatives of sulfate-reducing *Desulfotomaculum* sp., thiosulfate-reducing *Thermoanaerobacterium* sp., elemental sulfur-reducing and fermenting *Petrotoga* sp., and to uncultured bacteria found in, for example, geothermal water or petroleum reservoirs. A first 16S rRNA gene sequence analysis of the archaeal community of the formation water sample, which had been collected in situ, indicated the occurrence of members of hydrogenotrophic *Methanomicrobiales*.

The in situ-sampled formation water of a well bore of the "Salzwedel-Peckensen" subfield block was more diverse and comprised 16S rRNA gene sequences, which were assigned predominantly to uncultured bacteria detected in, for example, volcanic deposits, petroleum reservoirs, geothermal water, or hydrothermal vents. In addition, sequences have been affiliated with *Desulfotomaculum* sp., *Thermoanaerobacterium* sp., *Petrotoga* sp., and to *Delftia* sp. found in PAH-contaminated soils. Interestingly, *Desulfotomaculum* sp. has also been detected in the 16S rRNA sequence analysis of the living bacterial community of the same formation water sample [41].

Projects storing CO_2 in deep saline aquifers are, for example, Frio Brine Pilot Test (USA), CO_2SINK and CO_2MAN (Ketzin, Germany), Sleipner (Norway), and Nagaoka project (Japan). However, the deep subsurface biosphere has been considered only in CO_2SINK and CO_2MAN, two projects on the small-scale pilot CCS test site in Ketzin, Germany.

Since 2008, CO_2 has been injected ($\sim$60,000 t of mainly food-grade CO_2) into a saline aquifer, which is located in the "Roskow-Ketzin"double anticline, at a depth of 630–650 m below surface [53]. The CO_2 plume reached the first of the two observation wells two weeks after the start of CO_2 injection. The drill mud was removed from the injection well and the two observation wells using a N_2 lift at each well.

For microbial analyses, formation water of the first observation well has been collected in situ using either a flow-through sampler or a double-ball lining before and after CO_2 injection and at the well head during the N_2 lift. The microbial community has been analyzed using 16S rRNA gene fingerprinting methods (PCR-SSCP, DGGE) and cell counting with FISH and DAPI staining [70, 71]. Predominant microorganisms could be detected independently of the sampling procedure, which indicates negligible contamination effects during sampling. The microbial community was dominated by anaerobic halophilic fermentative bacteria (*Halanaerobium* sp., *Halobacteroidaceae*) and sulfate-reducing bacteria (*Desulfohalobium* sp., *Desulfotomaculum* sp.). Other members of the bacterial community were affiliated with phenanthrene-degrading *Comamonas* sp., to *Empedobacter* sp. from petroleum-oil contaminated soil and to oil-degrading bacteria of *Bacteroidetes*. After CO_2 arrival at the observation well, chemolithotrophic microorganisms temporarily outcompeted chemoorganotrophic microorganisms.

Microscopic analyses revealed total cell numbers of $2–6 \times 10^6$ and $2–4 \times 10^6$ cells/ml of living microorganisms before N_2 lift and CO_2 injection in formation water of the first observation well [70]. After N_2 lift, there were hardly any microorganisms detectable, but after CO_2 injection, total cell numbers were again determined to be 10^5 cells/ml. Moreover, after five months of CO_2 injection, total cell numbers again reached 2×10^6 cells/ml and comprised almost exclusively living microorganisms. Representatives of *Alpha-*, *Beta-* and *Gammaproteobacteria*, sulfate-reducing bacteria (*Desulfovibrionales*, *Desulfotomaculum* cluster I, and other *Firmicutes*, *Desulfobacteraceae*), and methanogenic archaea were detected using specific probes via FISH analyses.

Sulfate-reducing bacteria were detected in formation water of the injection well and were shown to be responsible for a decrease in the sulfate concentration and an increase in iron sulfide formation, which caused a decrease in permeability of the injection well and could be removed by a N_2 lift [102].

In addition, samples of drilling cores were investigated in long-term laboratory experiments with synthetic brine (172.8 g/l NaCl, 0.62 g/l KCl, 8.0 g/l $MgCl_2$ * $6H_2O$, 4.9 g/l $CaCl_2$ * $2H_2O$) under in situ conditions (5.5 MPa and 40 °C) and high CO_2 partial pressure to detect indigenous microorganisms and to quantify microbial activity [97]. The microbial community of the sandstone has been affiliated with members of *Alphaproteobacteria* (*Rhizobium* sp., *Agrobacterium* sp.), *Betaproteobacteria* (*Burkholderia* sp., *Hydrogenophaga* sp.), and *Actinobacteria* (*Propionibacterium* sp.). Except for *Agrobacterium* sp. and *Hydrogenophaga* sp., all other bacteria survived the exposure to CO_2. Sulfate-reducing bacteria and archaea were not detected in sandstone material. Mineral dissolution due to CO_2 exposure caused an increase in porosities during long-term experiments [96]. However, after 24 months, porosities again decreased due to precipitation [30].

1.4 Conclusion and Perspectives

Carbon capture and storage can be a fast-acting approach to mitigate CO_2 emissions and can provide a gain in time for the development of energy-efficient renewables. Hence, if all safety precautions were considered and a reasonable handling secured, CCS could contribute considerably to prevent climate change.

Enhanced gas and oil recovery using CO_2 or storage of natural gas are known, long-performed, and CCS-analogue approaches. Therefore, findings from these approaches can help to deduce geochemical and biogeochemical reactions in a CCS operation. Nevertheless, research on CCS depends on pilot and demonstration tests to gain detailed process knowledge.

An interdisciplinary approach combining geophysical, geochemical, and biogeochemical monitoring of the whole CCS operation (baseline, injection, long-term storage) will be required to understand complex processes in the storage site and to be able to react properly if any problems in operation occur. In this respect, determination of the baseline conditions, the original microbial community composition, and knowledge of process behavior are essential to predict and then prevent any failure in advance. For example, a decrease of injectivity has occurred as an immediate consequence of microbial activity and has been recovered with a N_2 lift at the CCS pilot plant near Ketzin (Germany). During the Frio Brine Pilot Test (USA), dissolved organic carbon (DOC) has increased by a factor of 100 from 1–5 to 500–600 mg/l after 20 days of CO_2 injection [48]. The organic carbon, mainly formate, acetate, and toluene, can probably be extracted by SC–CO_2 from the rock of the geological formation, but can also be a result of microbiological metabolism generating biomass and organic compounds.

During the RECOBIO-1 project 2005–2008, Ehinger et al. [26] already showed that microbial activity can have an impact on the performance of CCS in depleted natural gas fields. However, only four CCS pilot plant projects considered the deep subsurface biosphere in their monitoring concept to date. These are CO_2CRC Otway Basin project (Australia), CLEAN project (Altmark, Germany), and CO_2SINK and CO_2MAN projects (Ketzin, Germany). In the CO_2SINK project, recovery of microbial cell numbers and microbial activity was shown after CO_2 injection into the subsurface saline aquifer near Ketzin.

In addition to the CLEAN project, the deep subsurface biosphere in formation waters of well bores around the subfield block, which was formerly considered for CO_2 injection in the natural gasfield Altmark, was investigated in the RECOBIO-2 project 2008–2011.

In general, besides sulfate-reducing, metal-reducing, fermenting, and biocorrosive bacteria, many uncultured microorganisms have been detected by molecular genetic analyses. Cultivation of microorganisms of the deep subsurface is challenging due to low cell numbers, low microbial activity after sampling and extreme physicochemical requirements. However, cultivation approaches are required, because successful enrichment, isolation, and description of so far unknown microorganisms will further improve knowledge of biogeochemical processes.

Carbon capture and storage provides not only a possible measure to promote climate protection, but also valuable insights into subsurface environments.

2 Underground Gas Storage (Methane, Hydrogen) for Energy Generation

2.1 Introduction to Underground Gas Storage

The underground storage of natural gas has its origin in the beginning of the twentieth century when gas companies searched for a solution to balance out the seasonal fluctuation in the demand for gas used for space heating of buildings [16]. Currently, around 630 underground gas storage (UGS) facilities are in operation worldwide [29]. New interest in large-scale underground storage of energy has been sparked by the expanding renewable energy production worldwide. The increasing utilization of solar or wind sources [91, 92] leads to a high fluctuation in energy production, which can be adapted to the actual demand by using the electrical energy to form hydrogen or methane and subsequent storage of the gases. Large volumes of storage capacities are required for this issue, which most likely can be solved by underground storage [94].

Although underground gas storage has been standard for engineering for decades, the impact of microbial processes on underground gas storage has hardly been explored. An early example of the impact of microbial processes on underground gas storage is provided by an underground town-gas reservoir near Lobodice, Czech Republic, where conspicuous changes in the gas volume and composition have been observed during a seven-month period of gas storage in the 1980's. The gas volume decreased by 10–20 % in conjunction with an approximately 1.5-fold increase in the methane content and significant losses of hydrogen, carbon dioxide, and carbon monoxide. Cultivation of microbial communities present in the reservoir water and rocks revealed methanogenic archaea as drivers of the changes in stored town gas. Changes in the carbon-isotope signature of methane in the stored town gas supported the result [88]. As exemplified in this study, microorganisms living in the deep subsurface can have profound effects on underground gas storage with respect to gas loss and alteration of gas composition. The consequences of this gas alteration are discussed below in more detail.

2.2 Microbiology of Gas Storage Sites

Underground gas storage is performed in depleted gas or oil reservoirs, aquifers, and salt caverns. These reservoirs are characterized by temperatures above 35 °C with a temperature increase of ~3 °C per 100 m depth, high pressure (>7 MPa) [29], absence of oxygen, and high salinity. Microbial life is widespread in the crust of the earth and numerous mechanisms to deal with different environmental factors have evolved [79, 82]. Microorganisms have been isolated that withstand hydrostatic pressure of 100 MPa [90], salt concentrations of up to 300 g/l [76], or temperatures of 113 °C [10]. Therefore, UGS facilities cannot be considered simply as a geological formation with unique physicochemical characteristics, but need to be seen also as a microorganism habitat. Indeed, between 10^3–10^6 microorganisms per ml reservoir water have been recorded in porous rock reservoirs [29, 46, 88] and microbial life has also been proven in salt formations [95]. Here the questions arise how microorganisms live in such habitats, which factors control the microbial activities, and how the microbial processes affect the underground gas storage.

Free water is vital for microbial life so that the residual reservoir water serves as habitat for microorganisms. Microbial life in deep geological storage systems, such as oil and gas reservoirs, is controlled by the reservoir temperature, salinity, abundance of essential inorganic nutrients, and appropriate energy resources [60]. The temperature is generally seen as the limiting factor for the presence of living microorganisms, while the other factors control the size and activity of the microbial populations [40, 61]. Despite the documented growth at 113 °C of the Archaeon *Pyrolobus fumarii* [10], in situ observations indicate that microbial activity in oligotrophic reservoirs is restricted to temperatures below 80–90 °C [100]. Microorganisms gain their energy from complex electron transfer processes involving the oxidation of organic and inorganic compounds and subsequent reduction of a

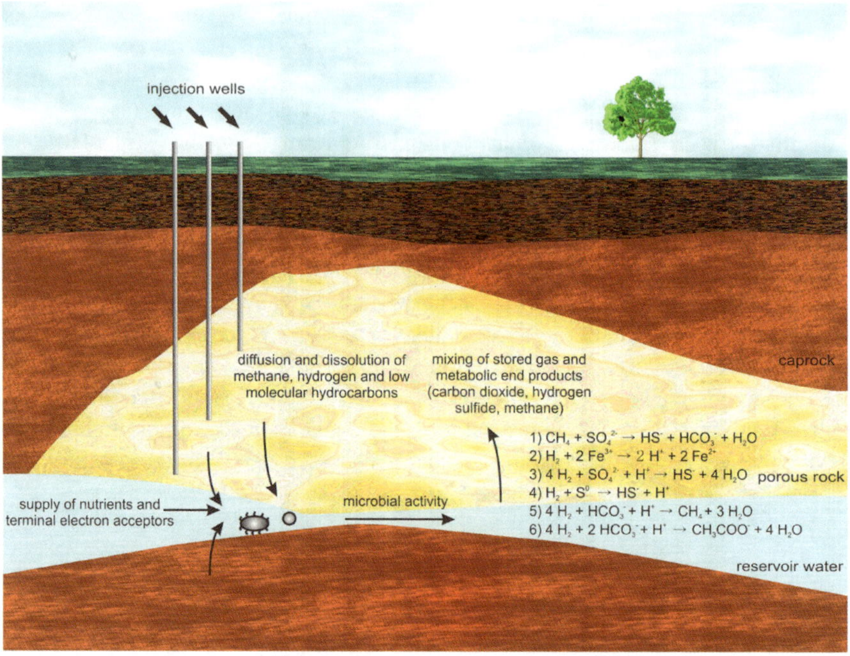

Fig. 3 Scheme of possible microbially mediated processes in underground gas reservoirs. Stored gas diffuses in the reservoir and dissolves in residual water, where gas components such as methane and hydrogen can be cosumed by microorganisms. Microorganisms derive nutrients such as phosphorous and nitrogen from mineral dissolution reactions, hydrolysis of dead cells, or from the nitrogen gas stored. Terminal electron acceptors are provided from marine evaporites, the mineral matrix, coal and shale layers, or also, in the case of carbon dioxide, from the stored gas itself. Volatile metabolic end products of microbial processes mix with the stored gas resulting in a change of the gas

terminal electron acceptor. The energy obtained is used for maintenance of microbial metabolism and growth [60]. Currently, methane and hydrogen are both considered as high-performance carriers of renewable energy, either directly produced in biogas plants or from the conversion of solar or wind energy. Both carriers will increasingly replace natural gas, which is a hydrocarbon mixture consisting primarily of methane and to a small extent of other low molecular hydrocarbons, carbon dioxide, nitrogen, and hydrogen sulfide [74] in both pipeline and storage systems. Hydrogen and methane as well as other low molecular hydrocarbons can serve as electron donors for microorganisms [40], so that the gas stored in a respective deep geological storage system provides sufficient energy sources for microbial activity (Fig. 3). Therefore, the depletion in essential inorganic nutrients, mainly phosphorous and nitrogen, and the availability of electron acceptors are considered as regulating factors for microbial activity. Suitable electron acceptors are ferric iron, manganese, sulfate, elemental sulfur, and carbon dioxide. Nitrate and nitrite are generally only present in low amounts [46, 60]. The electron acceptors are provided from the embedded or overlying marine evaporates, the mineral matrix (e.g., ferric iron containing siderite), coal, and shale layers or in case of carbon dioxide also from the stored gas itself. Nitrogen is present as ammonium ions in the water, which can be transported by reservoir water movements or diffusion or it can be assimilated from the nitrogen gas by nitrogen-assimilating microorganisms. Phosphor is considered as the much more likely limiting nutrient [40], which is present organically or inorganically bound and is mobilized by hydrolysis of dead cells or microbial-induced weathering of minerals such as phosphate-containing silicates [8, 83].

Methane, and with regard to future storage concepts, also hydrogen can be regarded as the dominant energy sources for microorganisms affecting the long-term fate of these stored gases. Although the solubility of both gases in water decreases with increasing temperatures and salinity [20, 101], the elevated pressure has a far greater impact on the solubility resulting in high dissolved gas concentrations in the water phase [5]. At elevated gas partial pressure, an increase in microbial activities has been recorded, which is attributable to the high availability of gaseous substrates in the water phase [22, 24, 55, 75]. In principle, anaerobic oxidation of methane (AOM) can proceed with sulfate as the terminal electron acceptor. The process is believed to be mediated by a syntrophic consortium of methanotrophic archaea and sulfate-reducing bacteria ([52] and references therein) or by methanotrophic archaea alone [65]. Furthermore, there are indications that methane oxidation is coupled with the reduction of manganese and ferric iron [6]. Thus far there is no single study addressing the role of AOM in gas reservoirs so that we can only speculate about its role.

Hydrogen plays a central role as an energy source in subsurface anoxic environments and can be utilized by a wide range of bacteria and archaea ([68, 89] and references therein). Hydrogen oxidation in such environments can be coupled to the reduction of ferric iron, sulfate, elemental sulfur, or carbon dioxide [19, 89]. Ferric iron reduction results in iron mobilization because the highly water-insoluble Fe(III) is reduced to the much more soluble Fe(II) [59]. Reduction of sulfate or elemental sulfur is highly undesirable in UGS because the formed hydrogen sulfide creates a serious problem for the industry due to its toxicity to humans [81], deterioration in quality, odor, souring, and corrosion of steel material of the well-tubing [7, 73, 93]. Moreover, hydrogen sulfide reacts with ferrous iron to form iron sulfide, which precipitates and can cause clogging of the operation equipment. Microbial mediated formation of hydrogen sulfide has been found repeatedly in gas and oil reservoirs [34, 80] pointing to the impact of sulfate reduction in UGS facilities. When all other electron acceptors are depleted methanogenesis and/or homoacetogenesis will appear, which are less favorable processes from a thermodynamic point of view [18]. In the course of methanogenesis, hydrogen oxidation is coupled to the reduction of carbon dioxide under formation of methane, a process which is exclusively mediated by archaea. Alternatively, homoacetogenic bacteria catabolize hydrogen and carbon dioxide to acetate. Both processes are widespread in the deep subsurface [54] and have been observed in gas reservoirs [46, 88]. In addition to the particular importance of methane and hydrogen as energy sources, microbial growth might also be stimulated by drilling fluids providing additional energy sources and nutrients as modeled by Baker [4]. This involves the risk of clogging of technical equipment by microbial biofilms or damage by microbial corrosion [11].

2.3 *Implications and Future Perspectives*

Overall, microbial activities lead to a loss of the stored gas, especially of hydrogen. Little is currently known about the extent. For example, in the course of methanogenesis, 4 mol hydrogen and 1 mol carbon dioxide are required to produce 1 mol methane and 3 mol water, which, for the operator of a UGS facility, means a substantial loss in the stored gas. Although the heating value of methane is with 35.9 MJ/m^3 higher than that of hydrogen (10.7 MJ/m^3), methanogenesis also means a loss in calorific power.

One may speculate that the highest microbial activity occurs near the gas–water contact where a plentiful supply of electron donors is given, but high microbial activity also occurs at the mineral–water contact. From shallow aquifers, it is known that sessile bacteria contribute over 90 % of the total bacterial community and only less than 10 % exist in the planktonic lifestyle [2, 37, 39]. The first cultivation experiments with samples from an underground town gas reservoir showed a much higher activity of methanogenic archaea using water and rocks from the reservoir compared to the sole use of water [88].

To summarize, the major microbial-induced risks associated with underground gas storage are (i) loss of the gas and thereby calorific loss; (ii) damage of technical equipment by biocorrosion and clogging through precipitates and biomass; and (iii) risk to operational safety and deterioration in quality by hydrogen sulfide formation. Therefore, the understanding of microbial activities in the deep underground is crucial for an economically successful operation of UGS. Microbiological studies are required to shed light on the identity of indigenous bacteria, their metabolism and activity, and factors controlling the type of microbial processes. This should be done in close cooperation with UGS operators, hydrogeologists, geologists, and chemists ensuring a comprehensive understanding of the complex processes in the deep subsurface.

3 Geothermal Energy Production

3.1 *Geothermal Energy*

Geothermal energy is the heat generated in the Earth. In geothermal plants, this energy is used as a source for heat supply ($T > 60$ °C) or to drive geothermal power plants ($T > 120$ °C).

The use of geothermal energy is generally subdivided into the operation in shallow depth (down to 400 m) and the operation in great depth (2,000–4,000 m). Shallow geothermal energy can be exploited principally worldwide and is often installed in private households for autonomous heat supply. Deep geothermal energy, on the other hand, is most efficient in regions where large temperature reservoirs exist (T > 100 °C), which are sufficient for electricity production. Worldwide, five countries use geothermal energy to produce around 20 % of their electricity (Costa Rica, El Salvador, Iceland, Kenya, and the Philippines) [9]. However, those are not the countries with the highest geothermal capacities. Even higher capacities are found in the United States of America (Table 1).

The productivity of a geothermal plant depends on a variety of factors, including the chemical composition of the thermal water, the water temperature, and the water production rate. Another important, yet rarely considered factor is the microbiology. In the subsurface, the majority of the microorganisms live attached to the rocks. However, microorganisms can also become detached and carried off with the produced thermal water and thus enter geothermal power plants. Therefore, the interaction of microbiological processes with geothermal plants should be considered from both sides. First, how do microorganisms influence the use of geothermal energy and second, how does the use of geothermal energy influence the subsurface microbiology (Fig. 4).

3.2 *Geothermal Energy and its Effects on Subsurface Microbiology*

3.2.1 Shallow Geothermal Energy

For the extraction of shallow geothermal energy, closed loop systems are installed (Fig. 4). The fluid inside the system extracts heat from the underground, which is used in different ways depending on the season: in winter, heat is extracted from the underground and used for heat supply of buildings; in summer, when the ambient temperature is higher than the underground temperature, the cold fluid is used to cool buildings. Subsequently, the warmed water is re-injected into the underground. As a consequence, the aquifer temperature range (of 10–12 °C) decreases and increases, respectively, and microorganisms will have to manage temperature fluctuations of ± 6 °C [14, 86].

Changes in temperature not only affect the metabolic activity of microorganisms, but also the composition of the overall microbial community. In summer, locally increased temperatures (e.g., at injections sites) can promote growth of mesophilic bacteria whereas heat extraction in winter promotes microbial species that grow at lower temperatures (psychrophilic microorganisms). Temperature fluctuations also affect the chemical composition of the groundwater as it changes the solubility of solids, liquids, and gases, including potential organic and

Table 1 Top 15 countries using geothermal energy

Geothermal Electricity Production		**Geothermal Direct Use**	
Country	GWh/yr	Country	GWh/yr
USA	16,603	China	20,932
Philippines	10,311	USA	15,710
Indonesia	9,600	Sweden	12,585
Mexico	7,047	Turkey	10,247
Italy	5,520	Japan	7,139
Iceland	4,597	Norway	7,000
New Zealand	4,055	Iceland	6,768
Japan	3,064	France	3,592
Kenya	1,430	Germany	3,546
El Salvador	1,422	Netherlands	2,972
Costa Rica	1,131	Italy	2,762
Turkey	490	Hungary	2,713
Papua New Guinea	450	New Zealand	2,654
Russia	441	Canada	2,465
Nicaragua	310	Finland	2,325

Source www.iea.org

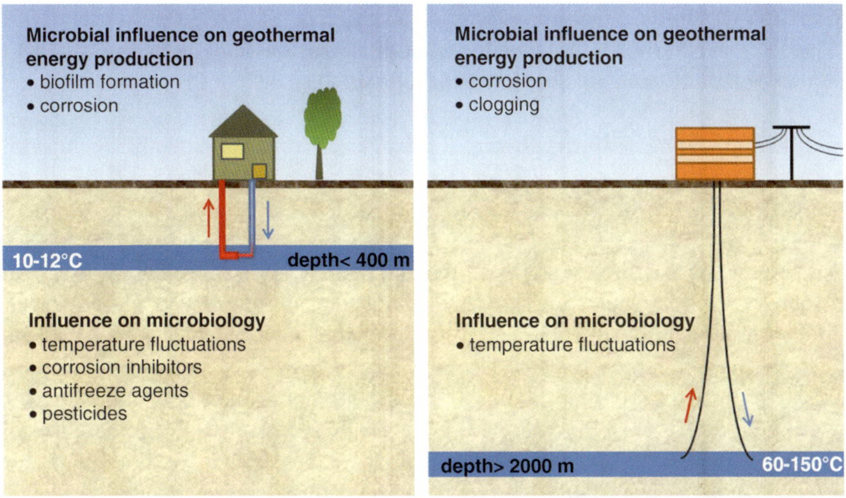

Fig. 4 Mutual influence of microbiology and geothermal energy in both shallow (*close loop, left*) and great (*open loop, right*) depth, exemplified for heat extraction from shallow and deep aquifers

inorganic substrates. Therefore, different substrate spectra will become available. For example, increased temperature will lower the solubility of oxygen (and other gases) and lead to a limitation of oxygen-dependent metabolic pathways.

3.2.2 Deep Geothermal Energy

Deep geothermal energy plants operate as open loop systems (Fig. 4) where hot water is extracted from the deep subsurface and is re-injected after passing the heat exchanger. The microbial community in greater depths considerably differs from that in shallow depth [57]. Despite the extreme conditions encountered in deep habitats (high temperatures and pressures, high salinity), deep aquifers have been shown to harbor a live and active biosphere [84]. Such ecosystems are often dominated by thermotolerant and thermophilic bacteria and archaea with mainly anaerobic metabolisms (e.g., fermenting, methanogenic, sulphate-reducing microorganisms) [1, 27, 71]. The deep subsurface also harbors populations of spore-forming bacteria, which are able to survive adverse conditions (e.g., heat, drought, substrate limitation) by formation of endospores. Metabolically, spores are largely inactive, but might germinate when temperature and nutrient supply conditions change and thereby influence the quality of geothermal water.

The operation of geothermal plants faces problems that mainly arise from the activity of sulfate-reducing bacteria. Sulfate reducers are capable of oxidizing iron ferrous metals, which results in corrosion of tubings and pipes [25, 28]. Also, formation of sulfidic precipitates (e.g., FeS) can lead to clogging and therefore to reduced water production rates [58]. Both corrosion and clogging can cause serious economic problems based on reduced performance of the geothermal plant.

3.3 Further Research

Temperature is an important factor that influences both microbial viability and metabolic activity. Therefore, research should focus on the effects of geothermal-induced temperature fluctuations on the chemical groundwater composition, nutrient supply, and the microbial community in both shallow and great depths.

Concerning the exploitation of shallow geothermal energy, potential leakage of fluid additives into groundwater raises questions concerning degradability and toxicity of the released substances, and the associated effects on microbial community composition. Also, the preservation of high groundwater quality is important because shallow groundwater is a source of drinking water (in Germany, 75 % is produced from it).

In the deep subsurface, most concerns arise from clogging and corrosion mediated by sulfate-reducing microorganisms. Sulfate reduction rates should be determined in order to estimate the extent of economic damage caused by these processes. Concerning spore-forming microbial populations, investigations are needed that address the effect of temperature fluctuations and/or changes in nutrient supply conditions on both the formation and germination of endospores.

References

1. Alawi M, Lerm S, Vetter A, Wolfgramm M, Seibt A, Würdemann H (2013) Diversity of sulfate-reducing bacteria in a plant using deep geothermal energy. Grundwasser 16(2):105–112
2. Alfreider A, Krössbacher M, Psenner R (1997) Groundwater samples do not reflect bacterial densities and activity in subsurface systems. Water Res 31(4):832–840
3. Arens P, Tuschewitzki GJ, Wollmann M, Follner H, Jacobi H (1995) Indicators for microbiologically induced corrosion of copper pipes in a cold-water plumbing system. Zentralbl Hyg Umweltmed 196(5):444–454
4. Baker SJ, West JM, Metcalfe R, Noy DJ, Yoshida H, Aoki K (1998) A biogeochemical assessment of the Tono site, Japan. J Contam Hydrol 35(1–3):331–340
5. Baranenko VI, Kirov VS (1989) Solubility of hydrogen in water in a broad temperature and pressure range. Sov Atom Energy 66(1):30–34
6. Beal EJ, House CH, Orphan VJ (2009) Manganese- and iron-dependent marine methane oxidation. Science 325(5937):184–187
7. Beech IB, Sunner J (2004) Biocorrosion: towards understanding interactions between biofilms and metals. Curr Opin Biotechnol 15(3):181–186
8. Bennett PC, Rogers JR, Choi WJ, Hiebert FK (2001) Silicates, silicate weathering, and microbial ecology. Geomicrobiol J 18(1):3–19

9. Bertani R (2007) World geothermal power generation in 2007. Presented at the European Geothermal Congress, Unterhaching, Germany, 30 May to 1 June 2007
10. Blöchl E, Rachel R, Burggraf S, Hafenbradl D, Jannasch HW, Stetter KO (1997) *Pyrolobus fumarii*, gen. and sp. nov., represents a novel group of archaea, extending the upper temperature limit for life to 113°C. Extremophiles 1(1):14–21
11. Bock M, Bosecker K, Kämpfer P, Dott W (1994) Isolation and characterization of heterotrophic, aerobic bacteria from oil storage caverns in northern Germany. Appl Microbiol Biotechnol 42(2–3):463–468
12. Boone DR, Liu Y, Zhao Z-J, Balkwill DL, Drake GR, Stevens TO, Aldrich HC (1995) *Bacillus infernus* sp. nov., an Fe(III)- and Mn(IV)-reducing anaerobe from the deep terrestrial subsurface. Int J Syst Bacteriol 45(3):441–448
13. Boreham C, Underschultz J, Stalker L, Kirste D, Freifeld B, Jenkins C, Ennis-King J (2011) Monitoring of CO_2 storage in a depleted natural gas reservoir: gas geochemistry from the CO_2CRC Otway Project, Australia. Int J Greenh Gas Con 5(4):1039–1054
14. Brielmann H, Lueders T, Schreglmann K, Ferraro F, Avramov M, Hammerl V, Blum P, Bayer P, Griebler C (2011) Oberflächennahe Geothermie und ihre potenziellen Auswirkungen auf Grundwasserökosysteme. Grundwasser 16(2):77–91
15. Bryant E (1997) Climate process and change. Cambridge University Press, Cambridge
16. Buschbach TC, Bond DC (1974) Underground storage of natural gas in Illinois, 1973. Department of Registration and Education, Illinois State Geological Survey, Urbana
17. Colwell FS, D'Hondt S (2013) Nature and extent of the deep biosphere. Rev Mineral Geochem 75:547–574
18. Conrad R (1999) Contribution of hydrogen to methane production and control of hydrogen concentrations in methanogenic soils and sediments. FEMS Microbiol Ecol 28(3):193–202
19. Cord-Ruwisch R, Seitz H-J, Conrad R (1988) The capacity of hydrogenotrophic anaerobic bacteria to compete for traces of hydrogen depends on the redox potential of the terminal electron acceptor. Arch Microbiol 149(4):350–357
20. Crozier TE, Yamamoto S (1974) Solubility of hydrogen in water, sea water, and sodium chloride solutions. J Chem Eng Data 19(3):242–244
21. Davis JB, Updegraff DM (1954) Microbiology in the petroleum industry. Bacteriol Rev 18(4):215
22. de Angelis MA, Baross JA, Lilley MD (1991) Enhanced microbial methane oxidation in water from a deep-sea hydrothermal vent field at simulated in situ hydrostatic pressures. Limnol Oceanogr 36(3):565–570
23. Decho AW (2010) Overview of biopolymer-induced mineralization: what goes on in biofilms? Ecol Eng 36(2):137–144
24. Deusner C, Meyer V, Ferdelman TG (2010) High-pressure systems for gas-phase free continuous incubation of enriched marine microbial communities performing anaerobic oxidation of methane. Biotechnol Bioeng 105(3):524–533
25. Dinh HT, Kuever J, Muszmann M, Hassel AW, Stratmann M, Widdel F (2004) Iron corrosion by novel anaerobic microorganisms. Nature 427(6977):829–832
26. Ehinger S (2011) Microbial communities and microbial processes in a German gas reservoir and their relevance for CO_2 storage. Dissertation, TU Bergakademie Freiberg, Freiberg
27. Ehinger S, Seifert J, Kassahun A, Schmalz L, Hoth N, Schlömann M (2009) Predominance of *Methanolobus* spp. and *Methanoculleus* spp. in the archaeal communities of saline gas field formation fluids. Geomicrobiol J 26(5):326–338
28. Enning D, Venzlaff H, Garrelfs J, Dinh HT, Meyer V, Mayrhofer K, Hassel AW, Stratmann M, Widdel F (2012) Marine sulfate-reducing bacteria cause serious corrosion of iron under electroconductive biogenic mineral crust. Environ Microbiol 14(7):1772–1787
29. Evans DJ (2008) An appraisal of underground gas storage technologies and incidents, for the development of risk assessment methodology. Report by the British Geological Survey to the Health & Safety Executive (HSE), HSE Research Report Series, Number RR605

30. Fischer S, Zemke K, Liebscher A, Wandrey M (2011) Petrophysical and petrochemical effects of long-term CO_2-exposure experiments on brine-saturated reservoir sandstone. Energy Procedia 4:4487–4494
31. Florette M, Rückheim J, Voigtländer G, Wendel H (2007) Gaz de France's current and future involvement in CCS projects—a commitment to sustainable development. Geotechnologien Sci Rep 9:73–81
32. Francis CA, Obraztsova AY, Tebo BM (2000) Dissimilatory metal reduction by the facultative anaerobe *Pantoea agglomerans* SP1. Appl Environ Microbiol 66(2):543–548
33. Gasda S, Bachu S, Celia M (2004) Spatial characterization of the location of potentially leaky wells penetrating a deep saline aquifer in a mature sedimentary basin. Environ Geol 46(6–7):707–720
34. Gieg LM, Jack TR, Foght JM (2011) Biological souring and mitigation in oil reservoirs. Appl Microbiol Biotechnol 92(2):263–282
35. Gniese C, Frerichs J, Krüger M, Hoth N, Kassahun A, Schlömann M (2011) The bacterial community in the natural gas reservoir Altmark (Germany) and the assessment of indigenous microorganisms. In: Biospektrum Sonderausgabe: Tagungsband zur VAAM-Jahrestagung 2011, Karlsruhe, Germany, 03–06 April 2011. Spektrum Akademischer Verlag, p 117
36. Graupner T, Kassahun A, Rammlmair D, Meima JA, Kock D, Furche M, Fiege A, Schippers A, Melcher F (2007) Formation of sequences of cemented layers and hardpans within sulfide-bearing mine tailings (mine district Freiberg, Germany). Appl Geochem 22(11):2486–2508
37. Griebler C, Mindl B, Slezak D, Geiger-Kaiser M (2002) Distribution patterns of attached and suspended bacteria in pristine and contaminated shallow aquifers studied with an in situ sediment exposure microcosm. Aquat Microb Ecol 28(2):117–129
38. Gunter WD, Perkins EH, McCann TJ (1993) Aquifer disposal of CO_2-rich gases: reaction design for added capacity. Energ Convers Manage 34(9–11):941–948
39. Hazen T, Jiménez L, López de Victoria G, Fliermans C (1991) Comparison of bacteria from deep subsurface sediment and adjacent groundwater. Microb Ecol 22(1):293–304
40. Head IM, Jones DM, Larter SR (2003) Biological activity in the deep subsurface and the origin of heavy oil. Nature 426(6964):344–352
41. Hoth N, Kassahun A, Krüger M, Gniese C, Frerichs J, Muschalle T, Reich M, Schlömann M (2011) Gesamtabschlussbericht des Forschungsprojekts: Untersuchung der biogeochemischen Transformation von im tiefen Untergrund gespeichertem CO_2, RECOBIO-2. Freiberg, Germany
42. Hoth N, Kassahun A, Seifert J, Krüger M, Bretschneider H, Gniese C, Frerichs J, Simon A, Simon E, Muschalle T (2009) Investigation of CO_2 induced biogeochemical reactions and active microorganisms of two German gas fields. Presented at the EGU General Assembly Conference, Wien, Austria, 19–24 April 2009
43. International Energy Agency Greenhouse Gas (IEAGHG) (2013) http://www.ieaghg.org/. Accessed 07 June 2013
44. IPCC (2005) IPCC special report on carbon dioxide capture and storage. In: Metz B, Davidson O, de Coninck HC, Loos M, Meyer LA (eds) Prepared by working group III of the intergovernmental panel on climate change. Cambridge University Press, Cambridge
45. IPCC (2007) Climate change 2007: the physical science basis. In: Solomon S, Qin D, Manning M, Chen Z, Marquis M, Averyt KB, Tignor M, Miller HL (eds) Contribution of working group I to the fourth assessment report of the intergovernmental panel on climate change. Cambridge University Press, Cambridge
46. Ivanova AE, Borzenkov IA, Tarasov AL, Milekhina EI, Belyaev SS (2007) A microbiological study of an underground gas storage in the process of gas injection. Microbiology 76(4):453–460
47. Jones DG, Lister TR, Smith DJ, West JM, Coombs P, Gadalia A, Brach M, Annunziatellis A, Lombardi S (2011) In Salah gas CO_2 storage JIP: surface gas and biological monitoring. Energy Procedia 4:3566–3573

48. Kharaka YK, Thordsen JJ, Hovorka SD, Seay Nance H, Cole DR, Phelps TJ, Knauss KG (2009) Potential environmental issues of CO_2 storage in deep saline aquifers: geochemical results from the Frio-I Brine Pilot test, Texas, USA. Appl Geochem 24(6):1106–1112
49. Kinrade SD, Del Nin JW, Schach AS, Sloan TA, Wilson KL, Knight CTG (1999) Stable five- and six-coordinated silicate anions in aqueous solution. Science 285(5433):1542–1545
50. Kleinitz W, Boehling E (2005) Underground gas storage in porous media-operating experience with bacteria on gas quality. Presented at the SPE Europec/EAGE Annual Conference, Madrid, Spain, 13–16 June 2005
51. Knauss KG, Johnson JW, Steefel CI (2005) Evaluation of the impact of CO_2, co-contaminant gas, aqueous fluid and reservoir rock interactions on the geologic sequestration of CO_2. Chem Geol 217(3–4):339–350
52. Knittel K, Boetius A (2009) Anaerobic oxidation of methane: progress with an unknown process. Annu Rev Microbiol 63(1):311–334
53. Köhler S, Zemke J, Becker W, Wiebach J, Liebscher A, Möller F, Bannach A (2013) Operational reservoir monitoring at the CO_2 pilot storage site Ketzin, Germany. In: Hou MZ, Xie H, Were P (eds) Clean energy systems in the subsurface: production, storage and conversion. Springer Series in Geomechanics and geoengineering. Springer, Berlin-Heidelberg, pp 53–63
54. Kotelnikova S (2002) Microbial production and oxidation of methane in deep subsurface. Earth-Sci Rev 58(3–4):367–395
55. Krüger M, Treude T, Wolters H, Nauhaus K, Boetius A (2005) Microbial methane turnover in different marine habitats. Palaeogeogr Palaeoclimatol Palaeoecol 227(1–3):6–17
56. Kühn M, Barth J, Baumann G, Becker V, Bock M, Buske S, Fritschen R, Giese R, Groß C, Henninges J, Houpt L, Kock D, Krüger M, Morozova D, Myrttinen A, Würdemann H (2013) Reservoir and cap rock monitoring. In: Kühn M, Münch U (eds) Clean. Advanced Technologies in Earth Sciences. Springer, Berlin-Heidelberg, pp 99–130
57. Lerm S, Alawi M, Miethling-Graff R, Seibt A, Wolfgramm M, Rauppack K, Würdemann H (2011) Mikrobiologisches Monitoring in zwei geothermisch genutzten Aquiferen des Norddeutschen Beckens. Z geol Wiss 39:195–212
58. Lerm S, Westphal A, Miethling-Graff R, Alawi M, Seibt A, Wolfgramm M, Würdemann H (2013) Thermal effects on microbial composition and microbiologically induced corrosion and mineral precipitation affecting operation of a geothermal plant in a deep saline aquifer. Extremophiles 17(2):311–327
59. Lovley DR (1991) Dissimilatory Fe(III) and Mn(IV) reduction. Microbiol Rev 55(2):259–287
60. Lovley DR, Chapelle FH (1995) Deep subsurface microbial processes. Rev Geophys 33(3):365–381
61. Magot M, Ollivier B, Patel BC (2000) Microbiology of petroleum reservoirs. Antonie Leeuwenhoek 77(2):103–116
62. Mahinpey N, Asghari K, Mirjafari P (2011) Biological sequestration of carbon dioxide in geological formations. Chem Eng Res Des 89(9):1873–1878
63. May F, Brune S, Gerling P, Krull P (2003) Möglichkeiten zur untertägigen Speicherung von CO_2 in Deutschland-eine Bestandsaufnahme. Geotechnik 26(3):162–172
64. Ménez B, Dupraz S, Gérard E, Guyot F, Rommevaux Jestin C, Libert M, Jullien M, Michel C, Delorme F, Battaglia-Brunet F (2007) Impact of the deep biosphere on CO_2 storage performance. Geotechnologien Sci Rep 9:150–163
65. Milucka J, Ferdelman TG, Polerecky L, Franzke D, Wegener G, Schmid M, Lieberwirth I, Wagner M, Widdel F, Kuypers MMM (2012) Zero-valent sulphur is a key intermediate in marine methane oxidation. Nature 491(7425):541–546
66. Mitchell AC, Phillips AJ, Hamilton MA, Gerlach R, Hollis WK, Kaszuba JP, Cunningham AB (2008) Resilience of planktonic and biofilm cultures to supercritical CO_2. J Supercrit Fluids 47(2):318–325

67. Mitchell AC, Phillips AJ, Hiebert R, Gerlach R, Spangler LH, Cunningham AB (2009) Biofilm enhanced geologic sequestration of supercritical CO_2. Int J Green Gas Con 3(1):90–99
68. Morita RY (1999) Is H_2 the universal energy source for long-term survival? Microb Ecol 38(4):307–320
69. Morozova D, Alawi M, Shaheed M, Krüger M, Kock D, Würdemann H (2011) The influence of microbial activity on rock fluid interaction: baseline characterization of deep biosphere for enhanced gas recovery in the Altmark natural gas reservoir. Energy Procedia 4:4633–4640
70. Morozova D, Wandrey M, Alawi M, Zimmer M, Vieth A, Zettlitzer M, Würdemann H (2010) Monitoring of the microbial community composition in saline aquifers during CO_2 storage by fluorescence in situ hybridisation. Int J Greenh Gas Con 4(6):981–989
71. Morozova D, Zettlitzer M, Let D, Würdemann H (2011) Monitoring of the microbial community composition in deep subsurface saline aquifers during CO_2 storage in Ketzin, Germany. Energy Procedia 4:4362–4370
72. Mu A, Billman-Jacobe H, Boreham C, Schacht U, Moreau JW (2011) How do deep saline aquifer microbial communities respond to supercritical CO_2 injection? Presented at the AGU Fall Meeting, San Francisco, California, 5–9 December 2011
73. Muyzer G, Stams AJM (2008) The ecology and biotechnology of sulphate-reducing bacteria. Nat Rev Micro 6(6):441–454
74. Natural Gas (2013) www.naturalgas.org. Accessed 07 June 2013
75. Nauhaus K, Boetius A, Krüger M, Widdel F (2002) In vitro demonstration of anaerobic oxidation of methane coupled to sulphate reduction in sediment from a marine gas hydrate area. Environ Microbiol 4(5):296–305
76. Oren A (2002) Diversity of halophilic microorganisms: environments, phylogeny, physiology, and applications. J Ind Microbiol Biotechnol 28(1):56–63
77. Palandri JL, Kharaka YK (2005) Ferric iron-bearing sediments as a mineral trap for CO_2 sequestration: iron reduction using sulfur-bearing waste gas. Chem Geol 217(3–4):351–364
78. Palandri JL, Rosenbauer RJ, Kharaka YK (2005) Ferric iron in sediments as a novel CO_2 mineral trap: CO_2–SO_2 reaction with hematite. Appl Geochem 20(11):2038–2048
79. Pikuta EV, Hoover RB, Tang J (2007) Microbial extremophiles at the limits of life. Crit Rev Microbiol 33(3):183–209
80. Raczkowski J, Turkiewicz A, Kapusta P (2004) Elimination of biogenic hydrogen sulfide in underground gas storage: a case study. Presented at the SPE Annual Technical Conference and Exhibition Houston, Texas, 26–29 September 2004
81. Reiffenstein RJ, Hulbert WC, Roth SH (1992) Toxicology of hydrogen sulfide. Annu Rev Pharmacol Toxicol 32(1):109–134
82. Reith F (2011) Life in the deep subsurface. Geology 39(3):287–288
83. Rogers JR, Bennett PC (2004) Mineral stimulation of subsurface microorganisms: release of limiting nutrients from silicates. Chem Geol 203(1–2):91–108
84. Sand W (2003) Microbial life in geothermal waters. Geothermics 32(4–6):655–667
85. Scherf A-K, Zetzl C, Smirnova I, Zettlitzer M, Vieth-Hillebrand A (2011) Mobilisation of organic compounds from reservoir rocks through the injection of CO_2—comparison of baseline characterization and laboratory experiments. Energy Procedia 4:4524–4531
86. Schippers A, Reichling J (2006) Laboruntersuchungen zum Einfluss von Temperaturveränderungen auf die Mikrobiologie des Untergrundes. Grundwasser 11(1):40–45
87. Scholes CA, Kentish SE, Stevens GW (2009) Effects of minor components in carbon dioxide capture using polymeric gas separation membranes. Sepa Purif Rev 38(1):1–44
88. Šmigáň P, Greksák M, Kozánková J, Buzek F, Onderka V, Wolf I (1990) Methanogenic bacteria as a key factor involved in changes of town gas stored in an underground reservoir. FEMS Microbiol Lett 73(3):221–224
89. Stevens T (1997) Lithoautotrophy in the subsurface. FEMS Microbiol Rev 20(3–4):327–337

90. Takami H, Inoue A, Fuji F, Horikoshi K (1997) Microbial flora in the deepest sea mud of the Mariana Trench. FEMS Microbiol Lett 152(2):279–285
91. The European Wind Energy Association (EWEA) (2013) Wind in power. 2012 European statistics
92. The World Wind Engery Association (WWEA) (2012) 2012—Half-year report
93. Videla HA, Herrera LK (2005) Microbiologically influenced corrosion: looking to the future. Int Microbiol 8(3):169–180
94. von Bremen L (2009) Storage and transport capacities in Europe for a full renewable power supply system. Presented at the European Wind Energy Conference, Marseille, France, 16–19 March 2009
95. Vreeland RH, Piselli AF Jr, McDonnough S, Meyers SS (1998) Distribution and diversity of halophilic bacteria in a subsurface salt formation. Extremophiles 2(3):321–331
96. Wandrey M, Fischer S, Zemke K, Liebscher A, Scherf A-K, Vieth-Hillebrand A, Zettlitzer M, Würdemann H (2010) Monitoring petrophysical, mineralogical, geochemical and microbiological effects of CO_2 exposure—Results of long-term experiments under in situ conditions. Energy Procedia 4:3644–3650
97. Wandrey M, Pellizari L, Zettlitzer M, Würdemann H (2011) Microbial community and inorganic fluid analysis during CO_2 storage within the frame of CO2SINK—Long-term experiments under in situ conditions. Energy Procedia 4:3651–3657
98. Wenzhöfer F, Asendorf V, Grünke S, Hagemann J, Hoge U, Hovland M, Lehmenhecker S, Shurn K, Weiz E, Wulff T (2012) Impact of potential CO_2 leakage on marine ecosystems—-RV Heincke cruise report HE 377, 16–24 April, Bremerhaven, Germany
99. White DC, Sutton SD, Ringelberg DB (1996) The genus *Sphingomonas*: physiology and ecology. Curr Opin Biotechnol 7(3):301–306
100. Wilhelms A, Larter SR, Head I, Farrimond P, di-Primio R, Zwach C (2001) Biodegradation of oil in uplifted basins prevented by deep-burial sterilization. Nature 411(6841):1034–1037
101. Yamamoto S, Alcauskas JB, Crozier TE (1976) Solubility of methane in distilled water and seawater. J Chem Eng Data 21(1):78–80
102. Zettlitzer M, Moeller F, Morozova D, Lokay P, Würdemann H (2010) Re-establishment of the proper injectivity of the CO_2-injection well Ktzi 201 in Ketzin, Germany. Int J Greenh Gas Con 4(6):952–959

Adv Biochem Eng Biotechnol (2014) 142: 123–146
DOI: 10.1007/10_2013_266

Published Online: 14 December 2013

Bioremediation via in situ Microbial Degradation of Organic Pollutants

Carsten Vogt and Hans Hermann Richnow

Abstract Contamination of soil and natural waters by organic pollutants is a global problem. The major organic pollutants of point sources are mineral oil, fuel components, and chlorinated hydrocarbons. Research from the last two decades discovered that most of these compounds are biodegradable under anoxic conditions. This has led to the rise of bioremediation strategies based on the in situ biodegradation of pollutants. Monitored natural attenuation is a concept by which a contaminated site is remediated by natural biodegradation; to evaluate such processes, a combination of chemical and microbiological methods are usually used. Compound specific stable isotope analysis emerged as a key method for detecting and quantifying in situ biodegradation. Natural attenuation processes can be initiated or accelerated by manipulating the environmental conditions to become favorable for indigenous pollutant degrading microbial communities or by adding externally breeded specific pollutant degrading microorganisms; these techniques are referred to as enhanced natural attenuation. Xenobiotic micropollutants, such as pesticides or pharmaceuticals, contaminate diffusively large areas in low concentrations; the biodegradation pattern of such contaminations are not yet understood.

Keywords In situ biodegradation · Micropollutants · Mineral oil · Monitored and enhanced natural attenuation

Abbreviations

BSS	Benzyl succinate synthase
BTEX	Benzene-toluene-ethylbenzene-xylenes
cDCE	cis-Dichloroethylene
CHC	Chlorinated hydrocarbons
CSIA	Compound-specific isotope analysis
2D-CSIA	Two-dimensional compound specific isotope analysis

C. Vogt (✉) · H. H. Richnow
Department of Isotope Biogeochemistry,
Helmholtz Centre for Environmental Research—UFZ,
Permoserstraße 15, 04318 Leipzig, Germany
e-mail: carsten.vogt@ufz.de

DNAPL	Dense nonaqueous phase liquids
EEA	European Environment Agency
ETBE	Ethyl-*tert*-butyl ether
ENA	Enhanced natural attenuation
HRC	Hydrogen releasing compounds
IRMS	Isotope ratio mass spectrometry
LC-IRMS	Liquid chromatography-isotope ratio mass spectrometry
mg/L	Milligram per liter
MNA	Monitored natural attenuation
MTBE	Methyl-*tert*-butyl ether
NAPL	Nonaqueous phase liquids
NRC	National Research Council
OSWER	Office of Solid Waste and Emergency Response
PAHs	Polycyclic aromatic hydrocarbons
PCE	Tetrachloroethylene or perchloroethylene
POPs	Persistent organic pollutants
TCE	Trichloroethylene
USA	United States of America
US-EPA	U.S. Environmental Protection Agency

Contents

1 Introduction: Distribution of Organic Pollutants 124
2 Biodegradation and in situ Bioremediation: Principles and Practice 128
  2.1 Biodegradation of Organic Pollutants: Principles 128
  2.2 Biodegradation of Organic Pollutants: A Short Overview 130
3 Bioremediation 133
  3.1 Monitored Natural Attenuation 134
  3.2 Enhanced Natural Attenuation 138
4 Conclusions and Outlook 140
References 140

1 Introduction: Distribution of Organic Pollutants

Environmental pollution by organic compounds is a major global problem. Due to the global use of a multitude of chemicals, urban and/or industrial areas are frequently contaminated by organics, forming polluted sites of different magnitudes [81]. In industrialized countries, pollution of soil and groundwater is mainly caused by industrial and commercial operations, municipal and industrial waste treatment, or inadequate storage of chemicals [18]; pollutants can originate from point sources such as waste pits, landfills, mine tailings, buried containers, or leaking storage tanks. In the USA, almost 300,000 polluted sites were reported [94]. Nearly

Table 1 Overview of main contaminants affecting soil in Europe [19]

Compound class	Frequency[a]
Heavy metals	37.3
Mineral oil	33.7
Polycyclic aromatic hydrocarbons (PAHs)	13.3
Aromatic hydrocarbons (BTEX)	6
Phenols	3.6
Chlorinated hydrocarbons	2.4
Others	3.6

[a] The ranking is calculated on the basis of the frequency (%) of contaminants reported to be the most important in the particular investigated site. The listed compound classes cover 100 % frequency

3 million sites are suspected to be polluted in Europe; around 250,000 of them require clean-up activities [18]. More than 20,000 sites are large-scale polluted megasites, usually characterized by multiple contaminant sources, complex contaminant cocktails, and high contaminant concentrations in soil and groundwater. The number of sites in Europe needing remediation may increase by 50 % by 2025, if trends of current investigations continue. In contrast, only 80,000 sites have been remediated in the last 30 years [18]. In Germany, more than 20,000 polluted sites have been identified by *Bund/Länder-Arbeitsgemeinschaft Boden-schutz* [52]; around 8,000 of them are currently remediated or monitored. Similar to the situation in Europe, the number of suspected contaminated sites exceeds the number of known contaminated sites by more than one order of magnitude [52].

Heavy metals and mineral oil are the main soil contaminants in Europe ([18]; Table 1). Important organic contaminants include polycyclic aromatic hydrocarbons (PAHs), aromatic hydrocarbons (benzene-toluene-ethylbenzene-xylenes, BTEX), phenols, and chlorinated hydrocarbons (CHC); globally, these compounds affect 90 % of contaminated sites, although their relative contributions may vary greatly from country to country [18]. Frequently detected contaminants in groundwater at polluted sites are chlorinated short-chain aliphatics, chlorinated aromatics, BTEX, and phenols (Table 2; [67, 68, 77]).

In contrast to point pollutions usually characterized by relatively large concentrations, organic contaminants are distributed in smaller concentrations over huge areas due to diffusive sources; such compounds have been termed *micropollutants* or emerging contaminants [81]. For example, pesticides used in agriculture can be widespread in low but potentially hazardous concentrations, causing chronic effects. The European Commission has listed 33 priority pollutants in the field of water policy; 11 of them are pesticides [25] (Table 3). Many pesticides belong to the category of persistent organic pollutants (POPs), which are toxic and resistant to environmental degradation. POPs are globally distributed by long-range transport processes and tend to bioaccumulate in human and animal tissue due to their high solubility in lipids. Thus, the production and usage of the most hazardous POPs has been globally limited or banned by the Stockholm Convention. The Stockholm Convention on Persistent Organic Pollutants is an international

Table 2 Frequently detected organic contaminants in groundwater [67, 68, 77] and general biodegradability

Chemical class	Compound	Applications	Biodegradation assessment
Highly chlorinated aliphatics	Tetrachloroethylen (PCE) trichloroethylen (TCE) 1,1,1-trichloroethane carbon tetrachloride	Dry cleaning fluids, degreasing solvents	Anoxic conditions, reductive dehalogenation: electron donor needed Slow or absent biodegradation under oxic conditions
Less chlorinated aliphatics	1,1-dichloroethane 1,2-dichloroethene vinyl chloride methylene chloride	Solvents, pesticides, landfills, biodegradation by-products, plastics	Oxic conditions; slow or absent biodegradation under anoxic conditions (electron donor needed here)
Less chlorinated aromatics	Chlorobenzene dichlorobenzenes	Solvents, industrial use	Oxic or anoxic conditions (electron acceptor needed here); chlorobenzene usually persistent under anoxic conditions
Low molecular weight hydrocarbons	BTEX, PAHs alkanes	Crude oil, refined fuels, dyestuffs, solvents, coal tar	Aerobic biodegradation faster than anaerobic biodegradation
Low molecular weight oxygenated hydrocarbons	Alcohols, ketones, esters, ethers, phenols	Solvents, paints, pesticides, adhesives, pharmaceuticals, fermentation products, detergents	Aerobic biodegradation faster than anaerobic biodegradation
	MTBE, ETBE	Fuel oxygenates	Aerobic biodegradation possible; very slow anaerobic biodegradation

environmental treaty effective since May 2004; it has been signed by 152 states. Currently, more than 20 POPs are listed or proposed to be listed by the Stockholm Convention ([87], Table 3).

Other more polar and water-soluble micropollutants such as drugs, hormones, or pharmaceuticals—compounds that are not completely removed during waste-water treatment and thus contaminate natural waters [81]—make up the group of emerging contaminants. Less persistent compounds, such as hormones or drugs, are biotically or abiotically transformed to various products, which can accumulate and can become even more toxic than the parent compounds [11]. Mixtures of low concentrated pollutants can be more toxic than single compounds due to synergistic effects, which are difficult to predict and thus complicate the evaluation of chemicals in the environment [3].

Table 3 List of organic priority substances or groups of organic priority substances in the field of water policy [25], and POPs listed by the Stockholm Convention (by May 2013)

Name of the substance	Origin	Usage, sources	Priority substance EC	Listed by Stockholm convention
Alachlor	Anthropogenic	Pesticide	X	
Aldrin	Anthropogenic	Pesticide		X
Anthracene, fluoranthene, naphthalene, other polyaromatic hydrocarbons	Natural	Coal tar, combustion	X	
Atrazine	Anthropogenic	Herbicide	X	
Benzene	Anthropogenic	Crude oil, gasoline, petrochemical	X	
Polybrominated diphenylether	Natural	Flame retardant	X	X
Cadmium and its compounds	Anthropogenic	Industry	X	
Chlordane	Anthropogenic	Pesticide		X
Chlordecone	Anthropogenic	Pesticide		X
Chlorinated naphthalines	Anthropogenic	Various industrial products		Proposed
Chlorinated paraffins (short-chained)	Anthropogenic	Various industrial products		Proposed
Chloroalkanes, C10-13	Anthropogenic	Chemical industry	X	
Chlorfenvinphos	Anthropogenic	Insecticide	X	
Chlorpyrifos	Anthropogenic	Insecticide	X	
DDT	Anthropogenic	Insecticide		X
1,2-Dichloroethane	Anthropogenic	Chemical industry	X	
Dichloromethane	Anthropogenic	Solvent	X	
Dieldrin	Anthropogenic	Insecticide		X
Bis (2-ethylhexyl) phthalate	Anthropogenic	Plasticizer	X	
Diuron	Anthropogenic	Herbicide	X	
Endrin	Anthropogenic	Insecticide		X
Endosulfan and its isomers	Anthropogenic	Insecticide	X	X
Heptachlor	Anthropogenic	Insecticide		X
Hexabromobiphenyl	Anthropogenic	Flame retardent		X
Hexabromocyclodecane	Anthropogenic	Flame retardent		X
Hexachlorobenzene	Anthropogenic	Fungicide	X	X
Hexachlorobutadiene	Anthropogenic	Solvent	X	Proposed
Hexachlorocyclohexane (α, β, γ-isomer)	Anthropogenic	Pesticide	X	X
Isoproturon	Anthropogenic	Herbicide	X	
Mirex	Anthropogenic	Insecticide		X

(continued)

Table 3 (continued)

Name of the substance	Origin	Usage, sources	Priority substance EC	Listed by Stockholm convention
Nonylphenols, octylphenols	Anthropogenic	Surfactant, industry	X	
Pentachlorobenzene	Anthropogenic	Chemical industry	X	X
Pentachlorophenol	Anthropogenic	Pesticide, desinfectant	X	Proposed
Perfluorooctane sulfonic acid (PFOS) and perfluorooctane sulfonyl fluoride (PFOS-F)	Anthropogenic	Various industrial products		X
Polychlorinated biphenyls (PCBs)	Anthropogenic	Industry		X
Polychlorinated dibenzofurans (PCDFs)	Anthropogenic	Industry		X
Polychlorinated dibenzo-p-dioxins (PCDDs)	Anthropogenic	Industry, combustion	X	X
Simazine	Anthropogenic	Herbicide	X	
Toxaphene	Anthropogenic	Insecticide		X
Tributyltin compounds	Anthropogenic	Biocide	X	
Trichlorobenzenes	Anthropogenic	Chemical industry	X	
Trichloromethane	Mainly natural	Chemical industry	X	

2 Biodegradation and in situ Bioremediation: Principles and Practice

2.1 Biodegradation of Organic Pollutants: Principles

Biodegradation has been defined as “the biologically catalyzed reduction in complexity of chemicals” [2]. This can lead to the conversion of carbon, nitrogen, phosphorous, sulfur, or other elements bound within the original organic chemical to inorganic products, which is named *mineralization*. If a compound is biologically transformed into products that are not metabolized or only very slowly further metabolized, a *transformation* has taken place. The transformation of hazardous chemicals into harmless chemicals is certainly a desirable biodegradation process, but sometimes the opposite is observed: a toxic compound is transformed into an even more toxic one, which accumulates in the environment. For example, the anaerobic dechlorination of higher chlorinated ethenes can lead to the accumulation of vinyl chloride, which is more toxic and volatile than its precursors [61]. Also, pesticides can be transformed to compounds that are more toxic than the parent molecule; the transformation of aldrin to dieldrin is one example [46].

Microorganisms are able to transform or mineralize almost every naturally occurring organic molecule, thereby driving the carbon cycle through organic matter decomposition. In contrast, synthetic anthropogenic chemicals containing nonbiotic functional groups or structures can be persistent because microorganisms have not evolved the biochemical machinery for degradation. Notably, anthropogenic compounds can be transformed or even mineralized in the environment, as some organisms have evolved enzymes with a rather broad substrate specificity. For example, the anthropogenic gasoline additives methyl *tert*-butyl ether (MTBE) or ethyl *tert*-butyl ether (ETBE) are widespread contaminants in the USA and Europe; they can form large contaminant plumes due to their high water solubility and recalcitrance to biodegradation. However, it has been observed that both MTBE and ETBE can be initially oxidized by several microorganisms using different monooxygenases with broad substrate spectra [57]. Meanwhile, a certain number of microorganisms were isolated, which were capable of mineralizing MTBE or ETBE while using it as a carbon and energy source (reviewed by Rosell et al. [75]).

A well-known enzyme performing many reactions is toluene dioxygenase: the enzyme from *Pseudomonas putida* strain F1 has been shown to catalyze at least 109 different reactions [92]. Sometimes, xenobiotics are even transformed by enzymes with a rather narrow substrate spectrum, as shown for the reductive dehalogenation of aliphatic and aromatic halogenated organics [89]. Additionally, due to their often short generation times and capabilities to change and exchange genetic information, microorganisms are able to evolve or acquire new degradation pathways in short time periods. This has been shown by the evolution of an aerobic chlorobenzene mineralizing strain at a chlorobenzene polluted site [101].

Generally, polluted sites are characterized by a cocktail of contaminants rather than a single compound. Contaminant cocktails are degraded by microbial communities consisting of several physiologically and/or phylogenetically different organisms. A particular species of that community typically mineralizes or transforms only a few constituents of the contaminant cocktail; therefore, degradation of the whole cocktail is due to the teamwork of a multitude of species. In anoxic systems, single contaminants are often mineralized by two or more species; if this behavior is necessary for the compound's mineralization, it is termed *syntrophy* [48]. Organic pollutant mineralization driven by syntrophic relationships is a significant process in highly reduced environments [85].

Besides being not biodegradable due to its physical–chemical properties, the persistence of a compound in a given environmental system can have different causes. A prerequisite for biodegradation is the presence of microorganisms that are actually able to degrade the compound. Several contaminant degraders might be ubiquitous, but not all. For example, aerobic benzene mineralizers are probably widespread, whereas anaerobic benzene mineralizers are not [106]. The contaminant must be also (bio)available for the degraders; organic molecules might be not bioavailable due to sorption to solids, the presence in nonaqueous phase lipids (NAPLs), or entrapment within the physical matrix of the environmental system [2]. Certain hydrophobic compounds, such as PAHs or higher molecular weight alipahic hydrocarbons, are not water-soluble and can strongly sorb to sediment particles. Such behavior can prevent, for example, a sustainable remediation of PAH-contaminated habitats [107]. The concentration of the contaminant in the environment is important: high concentrations might be toxic for the degrader population, whereas on the other hand a certain threshold concentration exist for each contaminant representing a level below the contaminant cannot be biodegraded [2]. The environmental conditions itself can support or impede biodegradation. Degraders are active in a certain temperature range; if the temperature is beyond this range, biodegradation will cease. Freshwater-adapted degraders might not be active in saline environments, and halophilic degraders are certainly inhibited in freshwater environments. Acidic or alkaline conditions will inhibit many degraders as they are adapted to neutral pH values. If a habitat is limited in a certain electron acceptor or nutrient, biodegradation might not occur, despite the presence of a suitable degrader community. In anoxic environments, the accumulation of fermentation metabolites can prevent the syntrophic degradation of a given contaminant by thermodynamic means; for example, this was described for benzene degradation at anoxic sites that were co-contaminated with ethanol [13]. In polluted aquifers, protists grazing on pollutant-degrading microbial communities can impact overall biodegradation rates [70].

A key environmental parameter for organic contaminant degradation is the availability of sufficient oxygen. Naturally occurring compounds and also several xenobiotics are usually quickly degraded or mineralized in the presence of air (aerobic conditions). In the case of high contaminant concentrations in aqueous systems, such as groundwater, oxygen is usually rapidly consumed by aerobic degraders, leading to prevalent anoxic conditions in the contaminated area. Until the late 1970s, it was hypothesized that most chemicals, even naturally occurring hydrocarbons, are not degradable under anoxic conditions [6]. Today, we assume that most chemicals can be transformed or mineralized in the absence of oxygen, although degradation rates, growth rates, and growth yields of anaerobic degraders are usually much lower than those of their aerobic counterparts. Several chemical compound classes, such as highly halogenated organics, are even preferably transformed under anoxic conditions [89]. Environmentally important electron acceptors for the anaerobic oxidation of organic molecules are sulfate, carbonate, iron, and nitrate.

In summary, detoxification of contaminated sites is mainly driven by microorganisms, and thus understanding the biochemistry and microbiology of contaminant-degrading microorganisms is essential for assessing natural attenuation processes.

2.2 Biodegradation of Organic Pollutants: A Short Overview

As outlined in the introduction, mineral oil components are major environmental pollutants. Mineral oil consists of more than 17,000 structurally different hydrocarbons and nitrogen, sulfur, or oxygen-containing derivatives [36]. Due to their

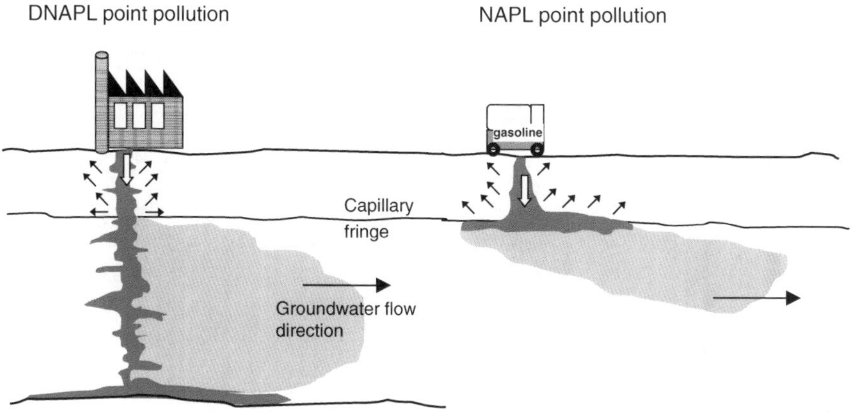

Fig. 1 Examples for point pollutions by DNAPLs and NAPLs. Modified from [101]

hydrophobic nature, mixtures of these compounds form NAPLs if present in high concentrations (illustrated in Fig. 1), complicating in situ biodegradation because compounds concentrated in NAPLs are poorly bioavailable. Nevertheless, mineral oil is generally biodegradable under a range of environmental conditions. Under aerobic conditions, such as in sea water, several mineral oil components can be quickly transformed and eventually mineralized if not limited by nutrients (particularly phosphorous and nitrogen) and bioavailability constraints [36]. In the presence of oxygen, hydrocarbons are usually activated by monooxygenases or dioxygenases, introducing one or two hydroxyl groups into the molecule [33, 54]; the formed alcohols are water soluble and thus bioavailable for further degradation steps. Ring cleavage of aromatics is as well catalyzed by oxygen-dependent dioxygenases [28, 33, 97]. Recent research results indicate that many oil components can be also mineralized under strictly anoxic conditions, although degradation rates are generally lower compared to aerobic conditions [42, 45, 108, 109]. In the absence of oxygen, hydrocarbons can be activated by carboxylation [10, 16, 65], hydroxylation [43, 49], or fumarate addition [37, 56]. Aromatic compounds are generally converted to benzoyl coenzyme A (benzoyl-CoA), which is further reduced by different benzoyl-CoA reductases [33, 71].

Prominent frequent and hazardous petroleum hydrocarbons are BTEX compounds, PAHs, and phenols due to their toxicity and global distribution; hence, anaerobic biodegradation of these compounds has been intensively studied in recent decades. BTEX compounds were shown to be mineralized under a range of different electron acceptor conditions (reviewed by Weelink et al. [105]). Field and laboratory data indicate that toluene is primarily degraded, followed by xylene, ethylbenzene, and benzene [5, 105]. Several anaerobic toluene degraders have been isolated; those activate toluene without exception by fumarate addition to the methyl moiety, producing benzyl succinate as the first intermediate. The enzyme catalyzing this reaction, benzylsuccinate synthase (BSS), has been used as a model enzyme for elucidating the biochemical mechanism of the fumarate addition reaction, which is widespread and used to activate several aliphatic and aromatic compounds [37, 108]. Also, all yet isolated anaerobic xylene degraders activate xylene isomers (*m*-xylene, *o*-xylene, *p*-xylene) by fumarate addition; notably, *m*-xylene is most readily degraded, whereas *p*-xylene is the worst degraded [105]. Ethylbenzene activation by fumarate addition has been also described for a marine sulfate reducer [50]. A few nitrate reducers were isolated, however, which initiate ethylbenzene degradation by an anaerobic hydroxylation step catalyzed by the enzyme ethylbenzene dehydrogenase [43, 49]. Benzene is the most toxic of the BTEX compounds: It is listed by the European Union as priority pollutant (Table 3) and is often persistent under anoxic conditions [5, 44]. Nevertheless, anaerobic benzene mineralization has been detected in the field and in laboratory cultures, mostly enrichment cultures reviewed by Vogt et al. [103]. It is currently not yet clear how benzene is activated in the absence of oxygen.

Among the PAHs, mineralization by stable enrichment cultures or pure cultures has been reported for naphthalene and phenanthrene under sulfate-reducing or iron-reducing conditions [16, 47, 63, 66, 110]. Cultures were achieved especially with naphthalene as substrate under sulfate- or iron-reducing conditions; growth was shown to be very slow in those cultures. It is not understood whether PAHs with four aromatic rings can be anaerobically mineralized; the results for naphthalene and phenanthrene indicate, however, that any mineralization might proceed tremendously slowly. Notably, high-molecular-weight polycyclic or heterocyclic aromatic compounds might be co-metabolically transformed by anaerobic aromatics degraders to more polar metabolites, such as by carboxylation reactions [76]. In different sulfate-reducing enrichment and pure cultures, PAH degradation was shown to be initiated by a carboxylation step [16, 65, 110].

Phenol and its derivatives differ considerably from pure hydrocarbons due to their generally higher water solubility, which makes them much more bioavailable. Phenols contain one or more reactive hydroxyl groups and were shown to be biodegraded under various electron acceptor conditions [71, 78]. Phenol is activated by an energy-dependent carboxylation step [9]. Notably, phenols are predominantly degraded by oxidative pathways under nitrate-reducing conditions and by reductive pathways under strictly anoxic conditions [71, 78].

Due to their common industrial applications, short-chained chlorinated aliphatics and chlorinated aromatics are frequent groundwater contaminants (Table 2). Furthermore, several POPs and pesticides are organochlorides (Table 3). Halogenated hydrocarbons are hydrophobic, poorly water-soluble, and heavier than water, thus forming dense nonaqueous phase liquids (DNAPLs) that sink in an aquifer (Fig. 1), reducing the bioavailability of the target compounds and hampering in situ biodegradation. The xenobiotic chlorinated organic substances listed in Tables 2 and 3 occurring in high concentration at industrial sites or being distributed on a large scale in low concentrations have natural analogues typically present in low concentrations in pristine environments; the huge number (more than 3,800; [34]) and global distribution of such naturally synthesized halogenated organic molecules is probably the key to understand why several microorganisms are able to degrade xenobiotic chlorinated compounds. Due to the high number of natural halogenated organics, microorganisms have evolved several different enzymatic systems transforming halogenated compounds [27]: oxidative dehalogenation catalyzed by mono- or dioxygenases, halide elimination, halide substitution, and reductive dehalogenation, which is the main reaction mechanism under anoxic conditions. The mentioned reactions can be metabolic or co-metabolic. As a rule of thumb, less chlorinated compounds are preferentially degraded under aerobic conditions, and higher chlorinated compounds are often degradable only under anoxic conditions [60, 84]. During reductive dehalogenation, the halogenated compound is used as an electron acceptor and not as a carbon source. Hydrogen is used by many reductive dehalogenating microorganisms as the preferred or even exclusive electron donor [57, 84], leading to a requirement of a hydrogen source from cooperating fermenting microorganisms. Because the potential energy release of reductive dehalogenation is considerably higher than that of sulfate reduction or methanogenesis [84], for example, reductive dehalogenating microorganisms can successfully compete for hydrogen in sulfate-reducing or methanogenic environments, which is an important aspect with regard to bioremediation. Members of the genus *Dehalococcoides* turned out to be key organisms for the reductive dechlorination of several widespread and toxic aliphatic and aromatic chlorinated compounds [1, 58, 89, 95]. *Dehalococcoides* were shown to dechlorinate even polychlorinated dibenzo-p-dioxins and dibenzofurans [12], which are compound classes containing acute toxic congeners [38].

Micropollutants are often not mineralized but transformed to various products that are more abundant than the parent compounds [11, 24, 81]. Transformation can occur metabolically in human or mammalian cells, during advanced wastewater treatment processes, or abiotically or biologically in the environment [24]. Generally, the in situ biodegradation patterns of micropollutants, such as pesticides, are currently not well understood [26]; for many of the priority substances listed in Table 3, the main in situ biodegradation pathways are not well understood.

3 Bioremediation

Bioremediation means the degradation of environmental pollutants by microorganisms. Generally, only point pollutions are bioremediated by specific technologies, as diffusively spread compounds are difficult to control due to their wide distribution and relatively low concentrations. Bioremediation technologies are usually separated as ex situ and in situ techniques. During ex situ bioremediation, the contaminated material is removed from the contaminated site by excavation (soil) or extraction techniques (groundwater, pump-and-treat) and treated elsewhere, such as in a bioreactor [53] or by composting. Ex situ bioremediation can be useful if soil or groundwater is contaminated by toxic organic compounds in areas threatening residents or sensitive ecosystems, or if toxic contaminants are insufficiently bioavailable, such as PAHs. Biodegradation of pollutants in excavated soil might be stimulated by inoculation with pollutant-degrading microorganisms or by the addition of nutrients or oxygen (e.g., [88]). However, ex situ bioremediation technologies have major shortcomings [68]. Ecosystems are destroyed by excavation of soil and sediments; furthermore, workers and nearby residents might be exposed to elevated levels of contaminants during excavation. Due to often unfavorable hydrological boundary conditions, not all contamination might be removed from a contaminated aquifer by pump-and-treat. In addition, both soil excavation and pump-and-treat are usually expensive and cost valuable resources.

As described above, research of the last two decades has indicated that many organic contaminants can be biologically transformed or mineralized under a range of environmental conditions. Therefore, in situ bioremediation approaches are increasingly being used for point source decontamination. In situ bioremediation can be divided in two classes: (i) technologies by which in situ biodegradation is actively enhanced, and (ii) the monitoring of ongoing in situ biodegradation without enhancement of natural attenuation processes. The former is termed enhanced natural attenuation (ENA), whereas the latter is termed monitored natural attenuation (MNA).

3.1 Monitored Natural Attenuation

The knowledge that the main organic pollutants are principally biodegradable (Table 2) led to the opinion that in situ bioremediation might be an option for many contaminated sites. The NRC of the USA stated in 1993 that in situ bioremediation is a scientifically valid and technically feasible technology if ongoing biodegradation reactions are clearly documented with several lines of evidence from the field [67]. Soon after this report, MNA was increasingly employed as a clean-up strategy at sites in the USA [68]. In 1999, the US-EPA defined the concept of monitored natural attenuation in the Office of Solid Waste and Emergency Response directive in more detail [93]. Notably, natural attenuation processes comprise per definition not only biodegradation and chemical degradation, but also a variety of physical or chemical mechanisms such as dispersion, dilution, sorption, volatilization, and (bio)chemical stabilization, which reduce the toxicity, mobility, or concentration of contaminants in soil and groundwater. However, biodegradation is considered to be the primary mechanism for the removal of the mass of contaminants [17]. In Germany between 2002 and 2008, the Federal Ministry of Education and Research funded several pilot studies regarding the effectiveness of MNA, resulting in a catalog of recommended procedures and monitoring methods [64]. Today, MNA is an accepted remediation strategy in many other European countries, although differences in application levels exist [17]. Nevertheless, due to the various microbiological and environmental parameters by which biodegradation rates can be influenced (outlined in Sect. 2.1), MNA is a concept that needs to be proven at every particular contaminated site. Typically, MNA lasts for some time, often several years to decades; thus, MNA is usually not an option for polluted sites characterized by short distance to humans, sensitive receptors, or sensitive ecosystems.

The keys for the successful operation of MNA at a site are the assessment of conceptual model of the flow and transport processes (hydrogeological model), and the use of reliable methods for verifying and quantifying in situ biodegradation processes. Several methods have been developed and used since the beginning of MNA operations (reviewed by Bombach et al. [10], Illmann and Alvarez [41], Scow and Hicks [83]): chemical methods such as geochemical approaches, tracer tests, metabolite analysis, compound-specific isotope analysis (CSIA), and microbiological methods such as in situ microcosms, detection of specific genes, and degradation tests using laboratory microcosms or columns. Generally, a combination of chemical and microbiological methods is used for proving natural attenuation processes at a specific site. Monitoring the pollutant concentrations in the source and the plume is always essential. Biogeochemical "footprints" may indicate in situ biodegradation of pollutants at the contaminated area, such as the consumption of electron acceptors (e.g., sulfate, nitrate), the accumulation of reduced electron acceptors (e.g., sulfide, methane), or the production of specific metabolites typical of a distinct pollution degradation pathway. The degradation potential of a site can be assessed by laboratory degradation tests with sediment or groundwater from the polluted site as well as the detection of genes coding for enzymes involved in specific pollutant degradation pathways. A more direct approach is the use of in situ microcosms, which enable proving the metabolization of ^{13}C-labelled target pollutants within the aquifer [86].

A novel noninvasive method, which combines qualitative and quantitative aspects of in situ biodegradation, is CSIA; due to its advantages, it has been increasingly used for proving natural attenuation processes, reflected also by a guideline published by the US-EPA [96]. CSIA exploits the fact that due to energetic constraints, stable light isotopes react usually slightly faster during biochemical transformation reactions than the corresponding stable heavy isotopes. As a consequence, residual specific organic contaminants get enriched in stable heavy isotopes (e.g., ^{13}C, ^{2}H, ^{37}Cl, ^{15}N) in the course of in situ biodegradation, a process termed *compound-specific stable isotope fractionation*, which can be analyzed by isotope ratio mass spectrometry (IRMS). The principles of CSIA in the context of in situ biodegradation have been described in several excellent reviews (e.g., [21, 39, 62, 80]).

Because physical removal processes such as dilution, sorption, or volatilization do not cause significant isotope fractionation, CSIA is a strong tool for proving in situ biodegradation. If the range of stable isotope fractionation of a specific biodegradation reaction has been determined by laboratory reference experiments with model microorganisms, CSIA may allow quantifying in situ biodegradation [29, 35, 90, 98, 99, 100]. In addition, by combining the analysis of two or more different stable isotopes within a molecule (two-dimensional CSIA), specific pathways of in situ biodegradation can be identified. For example, the strong hydrogen isotope fractionation linked to the anaerobic degradation of benzene leads

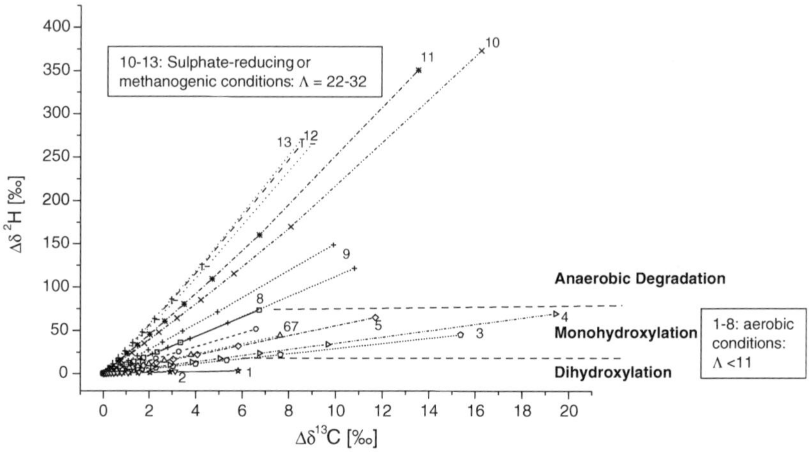

1) —★— *R. opacus* B-4, 2) --▽-- *P. putida* ML2, 3) ····○···· *Burkholderia* sp. (*26*), 4) ····▷···· *W. eutropha* 335, 5) --◇-- *A. denitrificans* strain BC (aerobic), 6) ····△···· *R. pickettii* PKO1, 7) --○-- *Acinetobacter* sp. (*26*), 8) —□— *A. denitrificans* strain BC (chlorate-reducing), 9) ····+···· Nitrate-reducing mixed culture (*29*), 10) ····×···· Sulphate-reducing mixed culture (*29*), 11) --✱-- Sulphate-reducing mixed culture (12 °C), 12) ····-···· Methanogenic mixed culture (*29*), 13) --I-- Sulphate-reducing mixed culture (20 °C)

Fig. 2 Plot of hydrogen versus carbon isotope fractionation (resulting in Λ values) for aerobic and anaerobic benzene biodegradation by different isolates or mixed cultures. Due to strong hydrogen isotope fractionation, the fractionation patterns for strictly anoxic cultures are significantly different from those from aerobic strains using mono- or dioxygenases for the initial benzene transformation reaction. Modified from [31]

to distinct dual hydrogen versus carbon isotope plots, allowing unambiguously detecting this reaction at benzene-contaminated field sites ([30–32]; Figs. 2 and 3). Isotope fractionation has been reported for transformation of most of the main organic contaminants listed in Table 2: BTEX, MTBE and ETBE, chlorobenzenes, chloroethenes, chloroethanes, or carbonchlorides (reviewed by Hunkeler and Morasch [40]), demonstrating the great potential of CSIA as a key MNA approach. Nevertheless, especially for aerobic biodegradation reactions catalyzed by mono- or dioxygenases, isotope fractionation can be below the detection limit; hence, the absence of any isotope fractionation at a field site is not a proof that a compound is not biodegraded. For example, the aerobic biodegradation of MTBE and ETBE can lead to either strong carbon and/or hydrogen isotope fractionation or to insignificant carbon and hydrogen isotope effects, depending on the type of enzyme (P450 or AlkB based, others) catalyzing the first reaction step of the degradation pathway [75]. Notably, this first step is similar for the different involved enzymes, a hydroxylation of the methyl group of the ether, showing that the simple relationship of "similar reaction = similar isotope pattern" does not exist; a similar observation has been made for the activation of toluene under anoxic conditions catalyzed by subtypes of the enzyme benzylsuccinate synthase [51, 104]. Thus, laboratory degradation experiments with reference cultures expressing distinct degradation pathways are a prerequisite for assessing biodegradation by CSIA at contaminated

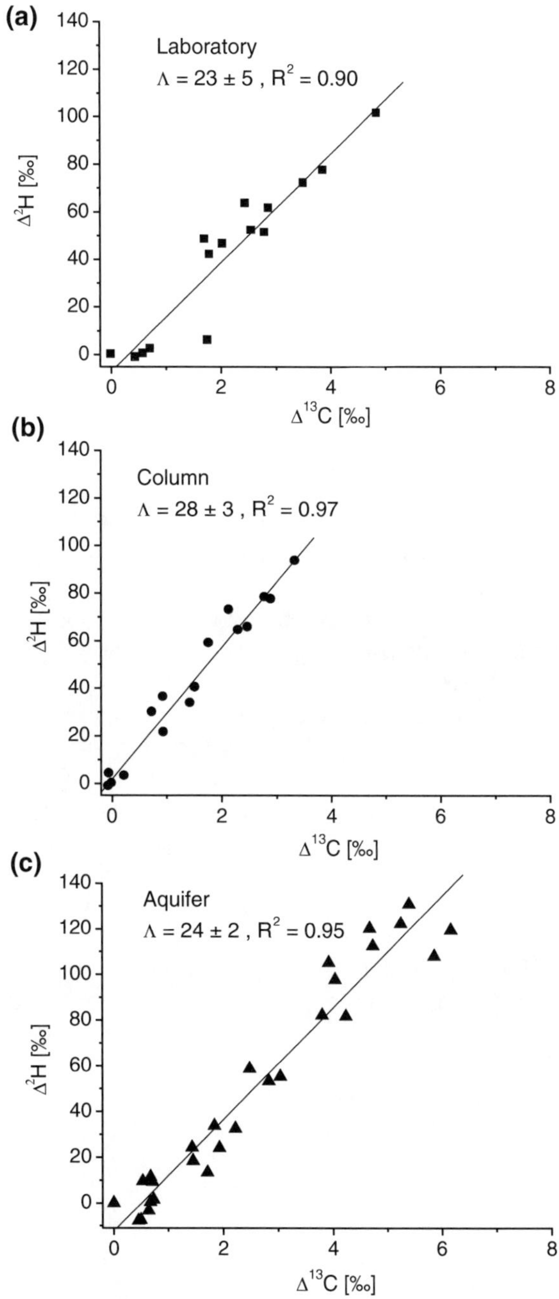

Fig. 3 Biodegradation-coupled carbon ($\Delta\ ^{13}C$) versus hydrogen ($\Delta\ ^{2}H$) isotope fractionation of benzene observed in **a** sulfate-reducing laboratory microcosms, **b** sulfate-reducing groundwater-percolated columns, and **c** aquifer samples. Laboratory microcosms and columns were prepared by using groundwater and/or sediment from the aquifer analyzed in (**c**). Experiments of different scales resulted in similar isotope fractionation patterns. Modified from [32]

sites. Besides characterizing in situ biodegradation, CSIA enables the detection and differentiation of contaminant sources [69, 79].

In summary, MNA is a reasonable technology if a site can be safely and sustainable remediated in an acceptable timeframe by natural attenuation processes. However, MNA often provides evidence that natural attenuation processes are too slow, slowed down, or even absent. In all these cases, stimulating natural attenuation processes by ENA is a reasonable option.

3.2 Enhanced Natural Attenuation

ENA takes advantage of stimulating in situ biodegradation by technical measures. Prior to any stimulation attempts, identification of bottlenecks and limitations of in situ biodegradation by MNA is essential. Depending on the type, mixture, and concentrations of pollutants and the site characteristics (e.g., hydrogeological conditions, geographic location, and age of plume), in situ biodegradation can be inhibited due to several reasons: limited availability of electron acceptors, donors, or nutrients; absence of biodegrading microbes; toxic or interfering pollutant or metabolite concentrations; and unfavorable pH values, salinity, temperatures, or redox conditions. Due to such different potential limitations, several ENA concepts have been attempted.

Because aerobic biodegradation is much faster than anaerobic biodegradation for many organic pollutants, several techniques have been developed to inject oxygen into the contaminated zone, even before MNA or ENA were accepted remediation concepts (e.g., [55]). These treatments were named differently, such as bio-venting, bio-sparging or air-sparging; typically, air, pure oxygen or hydrogen peroxide is used as an oxygen source. A major shortcoming of all these approaches is the low water solubility (approximately 10–15 mg/L under aquifer conditions) and the high chemical and microbiological reactivity of molecular oxygen, typically preventing a sustainable and fast biodegradation of organic pollutants in groundwater. Oxygen may not be solely used by microbes to activate and mineralize organic pollutants, but also for microbial and/or chemical oxidation of inorganics that are usually present in contaminant plumes, such as sulfide, ferrous iron, and others. As a consequence, oxygen must be supplied in much higher quantities than theoretically needed for the stoichiometric degradation of contaminants because all oxidation reactions beside contaminant degradation have to be taken into account. However, injecting air into contaminated groundwater can be reasonable if a pollutant is not or only very slowly degraded under anoxic conditions, but efficiently under microaerobic conditions, such as with chlorobenzene [7].

Often, injecting water-soluble salts that are usable for anaerobic pollutant degraders as electron acceptors, such as sulfates or nitrates, is an effective approach [4, 14]. Especially during in situ biodegradation of oil spills, the nutrients phosphorous and nitrogen may be limited due to significant growth of hydrocarbon-degraders; thus, the addition of nitrogen- and phosphorous-containing fertilizers can accelerate the degradation processes [91]. The remediation of oil spills might be also enhanced by adding biosurfactants or detergents, which generally improve the bioavailability of NAPL-forming oil components [91]. Such additions have to be handled with care, as detergents might increase the toxicity of the affected oil mixtures, as was shown for the detergent Corexit [74] which was massively applied during the Deepwater Horizon oil spill.

As outlined in Sect. 2.2, highly chlorinated compounds such as tetrachloroethene are reductively dechlorinated, whereby hydrogen is the most appropriate electron donor; also, acetate is beneficial for the growth of the reductive dechlorinating organisms. In practice, electron donors have been injected to contaminated field sites, which slowly released hydrogen and organics including acetate, turning the redox conditions to strongly reducing (e.g. polylactate ester HRC, emulsified vegetable oil, chitin, or biomass), finally leading to natural attenuation of the chlorinated target compounds [58].

Recently, the anaerobic degradation of BTEX was shown to be accelerated by the addition of ammonium acetate in a field experiment; acetate degradation resulted in strongly reducing conditions and promoted apparently the growth of anaerobic BTEX degraders [73], which might be a general strategy for enhancing anaerobic hydrocarbon degradation. On the other hand, acetate and hydrogen can inhibit the syntrophic degradation of hydrocarbons by thermodynamic means, as shown for benzene degradation [13, 15, 72]. Therefore, acetate and hydrocarbons are probably not simultaneously degraded, leading to lag-phases of hydrocarbon degradation in such approaches. Generally, biodegradation patterns of contaminant mixtures are difficult to predict due to possible negative or positive effects of the mixture itself upon the degrader microbial community. Furthermore, the accumulation of toxic metabolites or dead-end products can seriously affect enhanced natural attenuation processes. For example, oxygen injection into groundwater contaminated by aromatics or haloaromatics can lead to the accumulation of (halo)catechols due to fast ring oxidation by dioxygenases followed by slow subsequent metabolic oxidation reactions. Halocatechols are toxic because they can react with several biomolecules [82]; for 4-chlorocatechol, a dead-end enzymatic transformation to the antibiotic protoanemonin has been even reported [8].

It is possible that at certain locations, the number of specific pollutant degraders is too low to be significantly stimulated in acceptable time periods. Here, inoculating the contaminated site with specific degraders is a reasonable ENA concept, which is termed *bioaugmentation*. For example, at tetrachloroethylene or perchloroethylene-contaminated sites, *cis*-1,2-dichloroethene (cDCE) can accumulate as end product of reductive dechlorination due to the inactivity or absence of cDCE dechlorinating microorganisms. Reductive dechlorination of cDCE to nontoxic ethene—and hence successful bioremediation—was achieved by bioaugmentation with microbial cultures capable of this reaction [23, 59]. The success of bioaugmentation concepts depends on the "ecological robustness" and the mobility of the introduced microorganisms, as they have to compete with indigenous microbial communities and cope with site-specific hydrogeological conditions [20].

Generally, any bioaugmentation and biostimulation strategies require a deep understanding of the ecology of contaminant degradation at a specific site and need possibly a number of trials at the laboratory and pilot plant scale in preparation of a full field-scale remediation approach.

4 Conclusions and Outlook

The main organic pollutants—mineral oil and CHC—can be biodegraded by indigenous microorganisms. Research in recent decades has shown that especially anaerobic degradation reactions are driving natural attenuation processes at sites polluted by NAPLs or DNAPLs. Thus, MNA and ENA have become in situ remediation technologies in industrialized countries due to their noninvasive, sustainable, and cost-effective characteristics. Nevertheless, each contaminated site is unique with regard to hydrogeological conditions, contaminant cocktails, and potential hazards for humans and ecosystems, so that MNA and ENA should not be considered as standard practice per se. The analysis and quantification of processes controlling contaminant distribution in aquifer–soil systems require comprehensive site assessments, knowledge, and efforts. CSIA is a key monitoring method of MNA and ENA, as its application principally allows identification of distinct in situ degradation pathways, quantification of in situ biodegradation, and analysis of contaminant source apportionment. Multiple isotope analysis allows very precise elucidation of in situ biodegradation mechanisms. Recent and future method developments of stable isotope mass spectrometry will even extend the area of CSIA applications, such as chlorine and bromine stable isotope analysis of halogenated pollutants or liquid chromatography–IRMS based stable carbon analysis of polar organic pollutants. The development of CSIA applications for low-concentrated pesticides and other micropollutants is a challenging goal for future studies, although first steps in that direction have been taken [22]. The microbial degradation of micropollutants is currently not well understood, especially under anoxic conditions.

References

1. Adrian L, Szewzyk U, Wecke J, Görisch H (2000) Bacterial dehalorespiration with chlorinated benzenes. Nature 408:580–583
2. Alexander M (1994) Biodegradation and bioremediation. Academic Press, San Diego, California, USA
3. Altenburger R, Scholz S, Schmitt-Jansen M, Busch W, Escher BI (2012) Mixture toxicity revisited from a toxicogenomic perspective. Environ Sci Technol 46:2508–2522

4. Anderson RT, Lovley DR (2000) Anaerobic bioremediation of benzene under sulfate-reducing conditions in a petroleum-contaminated aquifer. Environ Sci Technol 34:2261–2266
5. Aronson D, Howard PH (1997) Anaerobic biodegradation of organic chemicals in groundwater: a summary of field and laboratory studies. Final Report. Environmental Science Center. Syracus Research Corporation, TR-97-02237
6. Atlas RM (1981) Microbial degradation of petroleum hydrocarbons: an environmental perspective. Microbiol Rev 45:180–209
7. Balcke GU, Paschke H, Vogt C, Schirmer M (2009) Pulsed gas injection: a minimum effort approach for enhanced natural attenuation of chlorobenzene in contaminated groundwater. Environ Pollut 157:2011–2018
8. Blasco R, Wittich RM, Mallavarapu M, Timmis KN, Pieper DH (1995) From xenobiotic to antibiotic, formation of protoanemonin from 4-chlorocatechol by enzymes of the 3-oxoadipate pathway. J Biol Chem 270:29229–29235
9. Boll M, Fuchs G (2005) Unusual reactions involved in anaerobic metabolism of phenolic compounds. Biol Chem 386:989–997
10. Bombach P, Richnow HH, Kästner M, Fischer A (2010) Current approaches for the assessment of in situ biodegradation. Appl Microbiol Biotechnol 86:839–852
11. Boxall AB, Sinclair CJ, Fenner K, Kolpin D, Maund SJ (2004) When synthetic chemicals degrade in the environment. Environ Sci Technol 38:368A–375A
12. Bunge M, Adrian L, Kraus A, Opel M, Lorenz WG, Andreesen JR, Görisch H, Lechner U (2003) Reductive dehalogenation of chlorinated dioxins by an anaerobic bacterium. Nature 421:357–360
13. Corseuil HX, Monier AL, Fernandes M, Schneider MR, Nunes CC, do Rosario M, Alvarez PJ (2011) BTEX plume dynamics following an ethanol blend release: geochemical footprint and thermodynamic constraints on natural attenuation. Environ Sci Technol 45:3422–3429
14. Cunningham JA, Rahme H, Hopkins GD, Lebron C, Reinhard M (2001) Enhanced in situ bioremediation of BTEX-contaminated groundwater by combined injection of nitrate and sulfate. Environ Sci Technol 35:1663–1670
15. da Silva ML, Gomez DE, Alvarez PJ (2013) Analytical model for BTEX natural attenuation in the presence of fuel ethanol and its anaerobic metabolite acetate. J Contam Hydrol 146:1–7
16. Davidova IA, Gieg LM, Duncan KE, Suflita JM (2007) Anaerobic phenanthrene mineralization by a carboxylating sulfate-reducing bacterial enrichment. ISME J 1:436–442
17. Declercq I, Cappuyns V, Duclos Y (2012) Monitored natural attenuation (MNA) of contaminated soils: state of the art in Europe–a critical evaluation. Sci Total Environ 426:393–405
18. EEA, European Environment Agency (2007) Progress in management of contaminated sites (CSI 015). http://www.eea.europa.eu/data-and-maps/indicators/progress-in-management-of-contaminated-sites/. Accessed 11 June 2013
19. EEA, European Environment Agency (2008) Overview of contaminants affecting soil and groundwater in Europe. http://www.eea.europa.eu/data-and-maps/figures/overview-of-contaminants-affecting-soil-and-groundwater-in-europe. Accessed 13 June 2013
20. El Fantroussi S, Agathos SN (2005) Is bioaugmentation a feasible strategy for pollutant removal and site remediation? Curr Opin Microbiol 8:268–275
21. Elsner M, Zwank L, Hunkeler D, Schwarzenbach RP (2005) A new concept linking observable stable isotope fractionation to transformation pathways of organic pollutants. Environ Sci Technol 39:6896–6916
22. Elsner M, Jochmann MA, Hofstetter TB, Hunkeler D, Bernstein A, Schmidt TC, Schimmelmann A (2012) Current challenges in compound-specific stable isotope analysis of environmental organic contaminants. Anal Bioanal Chem 403:2471–2491
23. Ellis DE, Lutz EJ, Odom JM (2000) Bioaugmentation for accelerated in situ anaerobic bioremediation. Environ Sci Technol 34:2254–2260

24. Escher BI, Fenner K (2011) Recent advances in environmental risk assessment of transformation products. Environ Sci Technol 45:3835–3847
25. European Commission (2008) Priority substances and certain other pollutants according to annex II of directive 2008/105/EC. http://ec.europa.eu/environment/water/water-framework/prioritysubstances.htm.11.06.2013. Accessed 13 June 2013
26. Fenner K, Canonica S, Wackett LP, Elsner M (2013) Evaluating pesticide degradation in the environment: blind spots and emerging opportunities. Science 341:752–758
27. Fetzner S (1998) Bacterial dehalogenation. Appl Microbiol Biotechnol 50:633–657
28. Fetzner S (2012) Ring-cleaving dioxygenases with a cupin fold. Appl Environ Microbiol 78:2505–2514
29. Fischer A, Bauer J, Meckenstock RU, Stichler W, Griebler C, Maloszewski P, Kästner M, Richnow HH (2006) A multitracer test proving the reliability of Rayleigh equation-based approach for assessing biodegradation in a BTEX contaminated aquifer. Environ Sci Technol 40:4245–4252
30. Fischer A, Theuerkorn K, Stelzer N, Gehre M, Thullner M, Richnow HH (2007) Applicability of stable isotope fractionation analysis for the characterization of benzene biodegradation in a BTEX-contaminated aquifer. Environ Sci Technol 41:3689–3696
31. Fischer A, Herklotz I, Herrmann S, Thullner M, Weelink SA, Stams AJ, Schlömann M, Richnow HH, Vogt C (2008) Combined carbon and hydrogen isotope fractionation investigations for elucidating benzene biodegradation pathways. Environ Sci Technol 42:4356–4363
32. Fischer A, Gehre M, Breitfeld J, Richnow HH, Vogt C (2009) Carbon and hydrogen isotope fractionation of benzene during biodegradation under sulfate-reducing conditions: a laboratory to field site approach. Rapid Commun Mass Spectrom 23:2439–2447
33. Fuchs G, Boll M, Heider J (2011) Microbial degradation of aromatic compounds—from one strategy to four. Nat Rev Microbiol 9:803–816
34. Gribble GW (2003) The diversity of naturally produced organohalogens. Chemosphere 52:289–297
35. Griebler C, Safinowski M, Vieth A, Richnow HH, Meckenstock RU (2004) Combined application of stable carbon isotope analysis and specific metabolites determination for assessing in situ degradation of aromatic hydrocarbons in a tar oil-contaminated aquifer. Environ Sci Technol 38:617–631
36. Head IM, Jones DM, Röling WF (2006) Marine microorganisms make a meal of oil. Nat Rev Microbiol 4:173–182
37. Heider J (2007) Adding handles to unhandy substrates: anaerobic hydrocarbon activation mechanisms. Curr Opin Chem Biol 1:188–194
38. Hites RA (2011) Dioxins: an overview and history. Environ Sci Technol 45:16–20
39. Hofstetter TB, Berg M (2011) Assessing transformation processes of organic contaminants by compound-specific stable isotope analysis. Trends Anal Chem 30:618–627
40. Hunkeler D, Morasch B (2010) Isotope fractionation during transformation processes. In: Aelion CM, Höhener P, Hunkeler D, Aravena R (eds) Environmental isotopes in biodegradation and bioremediation. CRC Press, Boca Raton, USA
41. Illmann WA, Alvarez PJ (2009) Performance assessment of bioremediation and natural attenuation. Crit Rev Env Sci Tech 39:209–270
42. Jiménez N, Morris BEL, Cai M, Yao J, Gründger F, Richnow HH, Krüger M (2012) Evidence for in situ methanogenic oil degradation in the Dagang oil field. Org Geochem 52:44–54
43. Johnson HA, Pelletier DA, Spormann AM (2001) Isolation and characterization of anaerobic ethylbenzene dehydrogenase, a novel Mo-Fe-S enzyme. J Bacteriol 183: 4536–4542
44. Johnson SJ, Woolhouse KJ, Prommer H, Barry DA, Christofi N (2003) Contribution of anaerobic microbial activity to natural attenuation of benzene in groundwater. Eng Geol 70:343–349

45. Jones DM, Head IM, Gray ND, Adams JJ, Rowan AK, Aitken CM, Bennett B, Huang H, Brown A, Bowler BF, Oldenburg T, Erdmann M, Larter SR (2008) Crude-oil biodegradation via methanogenesis in subsurface petroleum reservoirs. Nature 451:176–180
46. Jorgenson JL (2001) Aldrin and dieldrin: a review of research on their production, environmental deposition and fate, bioaccumulation, toxicology, and epidemiology in the United States. Environ Health Perspect 109:113–139
47. Kleemann R, Meckenstock RU (2011) Anaerobic naphthalene degradation by Gram-positive, iron-reducing bacteria. FEMS Microbiol Ecol 78:488–496
48. Kleinsteuber S, Schleinitz KM, Vogt C (2012) Key players and team play: anaerobic microbial communities in hydrocarbon-contaminated aquifers. Appl Microbiol Biotechnol 94:851–873
49. Kniemeyer O, Heider J (2001) Ethylbenzene dehydrogenase, a novel hydrocarbon-oxidizing molybdenum/iron-sulfur/heme enzyme. J Biol Chem 276:21381–21386
50. Kniemeyer O, Fischer T, Wilkes H, Glöckner FO, Widdel F (2003) Anaerobic degradation of ethylbenzene by a new type of marine sulfate-reducing bacterium. Appl Environ Microbiol 69:760–768
51. Kümmel S, Kuntze K, Vogt C, Boll M, Heider J, Richnow HH (2013) Evidence for benzylsuccinate synthase subtypes by stable isotope tools. J Bacteriol 195:4660–4667
52. LABO—Bund/Länder Arbeitsgemeinschaft Bodenschutz, Ständiger Ausschuss Altlasten (2012) Bericht des ALA über "Bundesweite Kennzahlen zur Altlastenstatistik" https://www.labo-deutschland.de/Veroeffentlichungen.html. Accessed 11 June 2013
53. Langwaldt JH, Puhakka JA (2000) On-site biological remediation of contaminated ground water: a review. Environ Pollut 107:187–197
54. Leahy JG, Batchelor PJ, Morcomb SM (2003) Evolution of the soluble diiron monooxygenases. FEMS Microbiol Rev 27:449–479
55. Lee MD, Thomas JM, Borden RC, Bedient PB, Ward CH, Wilson JT (1988) Biorestoration of aquifers contaminated with organic compounds. CRC Crit Rev Environ Control 18:29–87
56. Leuthner B, Leutwein C, Schulz H, Hörth P, Haehnel W, Schiltz E, Schaegger H, Heider J (1998) Biochemical and genetic characterization of benzylsuccinate synthase from *Thauera aromatica*: a new glycyl radical enzyme catalysing the first step in anaerobic toluene metabolism. Mol Microbiol 28:615–628
57. Lopes Ferreira N, Malandain C, Fayolle-Guichard F (2006) Enzymes and genes involved in the aerobic biodegradation of methyl *tert*-butyl ether (MTBE). Appl Microbiol Biotechnol 72:252–262
58. Löffler FE, Edwards EA (2006) Harnessing microbial activities for environmental cleanup. Curr Opin Biotechnol 17:274–284
59. Major DW, McMaster ML, Cox EE, Edwards EA, Dworatzek SM, Hendrickson ER, Starr MG, Payne JA, Buonamici LW (2002) Field demonstration of successful bioaugmentation to achieve dechlorination of tetrachloroethene to ethene. Environ Sci Technol 36:5106–5116
60. Maphosa F, Lieten SH, Dinkla I, Stams AJ, Smidt H, Fennell DE (2012) Ecogenomics of microbial communities in bioremediation of chlorinated contaminated sites. Front Microbiol 3:351. doi:10.3389/fmicb.2012.00351
61. Mattes TE, Alexander AK, Coleman NV (2010) Aerobic biodegradation of the chloroethenes: pathways, enzymes, ecology, and evolution. FEMS Microbiol Rev 34: 445–475
62. Meckenstock RU, Morasch B, Griebler C, Richnow HH (2004) Stable isotope fractionation analysis as a tool to monitor biodegradation in contaminated aquifers. J Contam Hydrol 75:215–255
63. Meckenstock RU, Mouttaki H (2011) Anaerobic degradation of non-substituted aromatic hydrocarbons. Curr Opin Biotechnol 22:406–414

64. Michels J, Stuhrmann M, Frey C, Koschitzky HP (2008) Handlungsempfehlungen mit Methodensammlung— Natürliche Schadstoffminderung bei der Sanierung von Altlasten. DECHEMA e.V., Frankfurt, Germany http://www.natural-attenuation.de/media/document/15_6948kora-handlungsempfehlungen.pdf. Accessed 30 July 2013
65. Mouttaki H, Johannes J, Meckenstock RU (2012) Identification of naphthalene carboxylase as a prototype for the anaerobic activation of non-substituted aromatic hydrocarbons. Environ Microbiol 14:2770–2774
66. Musat F, Galushko A, Jacob J, Widdel F, Kube M, Reinhardt R, Wilkes H, Schink B, Rabus R (2009) Anaerobic degradation of naphthalene and 2-methylnaphthalene by strains of marine sulfate-reducing bacteria. Environ Microbiol 11:209–219
67. National Research Council (1993) In situ bioremediation: when does it work? National Academies Press, Washington, USA
68. National Research Council (2000) Natural attenuation for groundwater remediation. National Academies Press, Washington, USA
69. Nijenhuis I, Schmidt M, Pellegatti E, Paramatti E, Richnow HH, Gargini A (2013) A combined stable isotope approach for source apportionment of chlorinated ethenes and related degradation processes at a complex multi-contamination event and plume field site. J Cont Hydrol (in press)
70. Novarino G, Warren A, Butler H, Lambourne G, Boxshall A, Bateman J, Kinner NE, Harvey RW, Mosse RA, Teltsch B (1997) Protistan communities in aquifers: a review. FEMS Microbiol Rev 20:261–275
71. Philipp B, Schink B (2012) Different strategies in anaerobic biodegradation of aromatic compounds: nitrate reducers versus strict anaerobes. Environ Microbiol Rep 4:469–478
72. Rakoczy J, Schleinitz KM, Müller N, Richnow HH, Vogt C (2011) Effects of hydrogen and acetate on benzene mineralisation under sulphate-reducing conditions. FEMS Microbiol Ecol 77:238–247
73. Ramos DT, da Silva ML, Chiaranda HS, Alvarez PJ, Corseuil HX (2013) Biostimulation of anaerobic BTEX biodegradation under fermentative methanogenic conditions at source-zone groundwater contaminated with a biodiesel blend (B20). Biodegradation 24:333–341
74. Rico-Martínez R, Snell TW, Shearer TL (2013) Synergistic toxicity of Macondo crude oil and dispersant Corexit 9500A(®) to the *Brachionus plicatilis* species complex (Rotifera). Environ Pollut 173:5–10
75. Rosell M, Gonzalez-Olmos R, Rohwerder T, Rusevova K, Georgi A, Kopinke FD, Richnow HH (2012) Critical evaluation of the 2D-CSIA scheme for distinguishing fuel oxygenate degradation reaction mechanisms. Environ Sci Technol 46:4757–4766
76. Safinowski M, Griebler C, Meckenstock RU (2006) Anaerobic cometabolic transformation of polycyclic and heterocyclic aromatic hydrocarbons: evidence from laboratory and field studies. Environ Sci Technol 40:4165–4173
77. Schiedek T, Grathwohl P, Teutsch G (1997) Literature study regarding Natural Attenuation/biodegradation of pollutants in groundwater. University of Tübingen, Chair of Applied Geology
78. Schink B, Philipp B, Müller J (2000) Anaerobic degradation of phenolic compounds. Naturwissenschaften 87:12–23
79. Schmidt TC, Jochmann MA (2012) Origin and fate of organic compounds in water: characterization by compound-specific stable isotope analysis. Annu Rev Anal Chem 5:133–155
80. Schmidt TC, Zwank L, Elsner M, Berg M, Meckenstock RU, Haderlein SB (2004) Compound-specific stable isotope analysis of organic contaminants in natural environments: a critical review of the state of the art, prospects, and future challenges. Anal Bioanal Chem 378:283–300
81. Schwarzenbach RP, Escher BI, Fenner K, Hofstetter TB, Johnson CA, von Gunten U, Wehrli B (2006) The challenge of micropollutants in aquatic systems. Science 313:1072–1077

82. Schweigert N, Zehnder AJB, Eggen RIL (2001) Chemical properties of catechols and their molecular modes of toxic action in cells, from microorganisms to mammals. Environ Microbiol 3:81–91
83. Scow KM, Hicks KA (2005) Natural attenuation and enhanced bioremediation of organic contaminants in groundwater. Curr Opin Biotechnol 16:246–253
84. Smidt H, de Vos WM (2004) Anaerobic microbial dehalogenation. Annu Rev Microbiol 58:43–73
85. Stams AJ, Plugge CM (2009) Electron transfer in syntrophic communities of anaerobic bacteria and archaea. Nat Rev Microbiol 7:568–577
86. Stelzer N, Buning C, Pfeifer F, Dohrmann AB, Tebbe CC, Nijenhuis I, Kästner M, Richnow HH (2006) In situ microcosms to evaluate natural attenuation potentials in contaminated aquifers. Org Geochem 37:1394–1410
87. Stockholm Convention (2013) The 12 Initial POPs, http://chm.pops.int/Convention/ThePOPs/The12InitialPOPs/tabid/296/Default.aspx; The New POPs, http://chm.pops.int/Convention/ThePOPs/TheNewPOPs/tabid/2511/Default.aspx. Accessed 14 June 2013
88. Sun GD, Xu Y, Jin JH, Zhong ZP, Liu Y, Luo M, Liu ZP (2012) Pilot scale ex-situ bioremediation of heavily PAHs-contaminated soil by indigenous microorganisms and bioaugmentation by a PAHs-degrading and bioemulsifier-producing strain. J Hazard Mater 233–234:72–78
89. Taş N, van Eekert MH, de Vos WM, Smidt H (2010) The little bacteria that can—diversity, genomics and ecophysiology of '*Dehalococcoides*' spp. in contaminated environments. Microb Biotechnol 3:389–402
90. Thullner M, Fischer A, Richnow HH, Wick LY (2013) Influence of mass transfer on stable isotope fractionation. Appl Microbiol Biotechnol 97:441–452
91. Tyagi M, da Fonseca MM, de Carvalho CC (2011) Bioaugmentation and biostimulation strategies to improve the effectiveness of bioremediation processes. Biodegradation 22:231–241
92. University of Minnesota (2013) Biocatalysis/Biodegradation database. http://umbbd.ethz.ch/tol/tdo.html. Accessed 27 June 2013
93. United States Environmental Protection Agency (US-EPA) (1999) Use of monitored natural attenuation at superfund, RCRA corrective action, and underground storage tank sites—directive OSWER 9200, pp 4–17
94. United States Environmental Protection Agency (US-EPA) (2004) Cleaning up the nation's waste sites: markets and technology trends. Office of Solid Waste and Emergency Response, Washington DC
95. United States Environmental Protection Agency (US-EPA) (2006) Evaluation of the role of *Dehalococcoides* in the natural attenuation of chlorinated ethylenes in ground water. Office of Research and Development, Ada, Oklahoma
96. United States Environmental Protection Agency (US-EPA) (2008) A guide for assessing biodegradation and source identification of organic ground water contaminants using compound specific isotope analysis (CSIA). Office of Research and Development, Ada, Oklahoma
97. Vaillancourt FH, Bolin JT, Eltis LD (2006) The ins and outs of ring-cleaving dioxygenases. Crit Rev Biochem Mol Biol 41:241–267
98. van Breukelen BM, Prommer H (2008) Beyond the Rayleigh equation: reactive transport modeling of isotope fractionation effects to improve quantification of biodegradation. Environ Sci Technol 42:2457–2463
99. van Breukelen BM (2007) Extending the Rayleigh equation to allow competing isotope fractionating pathways to improve quantification of biodegradation. Environ Sci Technol 41:4004–4010
100. van Breukelen BM (2007) Quantifying the degradation and dilution contribution to natural attenuation of contaminants by means of an open system rayleigh equation. Environ Sci Technol 41:4980–4985

101. van der Meer JR, Werlen C, Nishino SF, Spain JC (1998) Evolution of a pathway for chlorobenzene metabolism leads to natural attenuation in contaminated groundwater. Appl Environ Microbiol 64:4185–4193
102. Vogt C, Alfreider A, Griebler C (2003) Grundwasserverschmutzung und—sanierung. In: Griebler C, Mößlacher F (eds) Grundwasser-Ökologie. Eine Einführung. Uni-Taschen bücher (UTB) GmbH, vol 2111, pp 367–408
103. Vogt C, Kleinsteuber S, Richnow HH (2011) Anaerobic benzene degradation by bacteria. Microb Biotechnol 4:710–724
104. Vogt C, Cyrus E, Herklotz I, Herrmann S, Bahr A, Richnow HH, Fischer A (2008) Evaluation of aerobic and anaerobic toluene degradation pathways by two dimensional stable isotope fractionation. Environ Sci Technol 42:7793–7800
105. Weelink SAB, van Eekert MHA, Stams AJM (2010) Degradation of BTEX by anaerobic bacteria: physiology and application. Rev Environ Sci Biotechnol 9:359–385
106. Weiner JM, Lovley DR (1998) Anaerobic benzene degradation in petroleum-contaminated aquifer sediments after inoculation with a benzene-oxidizing enrichment. Appl Environ Microbiol 64:775–778
107. Weissenfels WD, Klewer HJ, Langhoff J (1992) Adsorption of polycyclic aromatic hydrocarbons (PAHs) by soil particles: influence on biodegradability and biotoxicity. Appl Microbiol Biotechnol 36:689–696
108. Widdel F, Rabus R (2001) Anaerobic biodegradation of saturated and aromatic hydrocarbons. Curr Opin Biotechnol 12:259–276
109. Zengler K, Richnow HH, Rossello-Mora R, Michaelis W, Widdel F (1999) Methane formation from long-chain alkanes by anaerobic microorganisms. Nature 401:266–269
110. Zhang X, Young LY (1997) Carboxylation as an initial reaction in the anaerobic metabolism of naphthalene and phenanthrene by sulfidogenic consortia. Appl Environ Microbiol 63:4759–4764

Adv Biochem Eng Biotechnol (2014) 142: 147–167
DOI: 10.1007/10_2013_178

Published Online: 11 April 2013

The Microbial Desulfurization of Coal

Giovanni Rossi

Abstract The chemical structure of coal macerals is usually characterized by the presence of inorganic and organic sulfur. Inorganic sulfur consists mostly of iron sulfides, the so-called "pyritic sulfur," whereas organic sulfur is covalently bound to the carbon atoms of the coal macromolecule. Comminution of coal to sizes that liberate the iron sulfide grains makes their removal with mineral beneficiation processes theoretically possible, but practically profitless. Microbial removal of pyritic sulfur has been extensively investigated over the last 50 years and the very promising results obtained have encouraged the design and construction of a semi-commercial pilot plant in the framework of Project JOULE 0039 funded by the European Commission. The results of the 1-year operation of this plant are reported here, the most significant being the 90 % pyrite removal achieved in five stirred tank bioreactors operating with a 40 % solids suspension and the pyritic iron solubilization rate of 36 mg dm^{-3} h^{-1}. Taking into account the very high price of the kWh in Italy, a rough estimate of the overall costs is in the range from 25 to 30 € per tonne of dry coal. So far the development of a microbial process for organic sulfur removal has shown to be much more difficult and less successful, although significant progress in laboratory research is reported.

Keywords *Acidithiobacillus ferrooxidans* · *Alcaligenes denitrificans* · *Brevibacterium* sp. · Coal · cos φ · Dibenzothiophene · EPS · Iron sulfides · *Leptospirillum ferrooxidans* · Metabolic pathway · Organic sulfur · Pachuca reactor · Pilot plant · *Pseudomonas* sp. · Pyritic sulfur · Sulfur

Abbreviations

C.E.C	Commission of European Communities
DBT	Dibenzothiophene
D.M.T	Deutsche Montan Technologie
E.C	European Community
E.N.I	Ente Nazionale Idrocarburi
EPS	Extracellular polymer substance
EU	European Union
m.o.g	Mesh-of-grind

G. Rossi (✉)
Geoengineering and Environmental Technologies Department,
University of Cagliari, Cagliari, Italy
e-mail: anralaba@alice.it

Contents

1 Chemical and Physical Occurrence of Sulfur in Coal 148
  1.1 The Removal of Pyritic Sulfur 150
  1.2 The Removal of Organic Sulfur 153
2 Semi-Commercial Coal Biodepyritization Operation 156
  2.1 The Comminution Bay 158
  2.2 The Bioreactor Bay, Depyritized Coal Stockpile, and Reject Water Disposal 160
3 Concluding Remarks and Outlook 162
References 163

1 Chemical and Physical Occurrence of Sulfur in Coal

Much work has been carried out by researchers worldwide to elucidate the chemical structure of the so-called "coal macerals" which is the name commonly given by petrographers to the complex organic materials occurring in coal in various petrographic types. This research—mostly related to coal liquefaction—was (and probably still is) aimed mainly at the identification of the macromolecules forming the macerals, also with a view to characterizing the form of the foreign matter contained therein, with special regard to sulfur.

Thus, there is currently complete consensus that sulfur occurs in coal in two forms: "inorganic" or "pyritic" and "organic" sulfur. Actually, minor amounts of sulfur can sometimes occur as sulfate sulfur as the result of oxidation of the inorganic sulfur, or even as elemental sulfur [5, 24].

The inorganic sulfur occurs mainly in the form of iron sulfides [47], by far the most frequent ones being pyrite and marcasite (altogether >90 %). An exhaustive review on the compounds of iron and sulfur can be profitably consulted [59]. According to a recent publication [22], less than 0.05 % of total sulfur in coal is present as sulfates. total sulfur and sulfur form distribution is often variable [11]. However, The possible compounds that can be formed by sulfur and iron are shown in Table 1 [59, 61–63]. They are crystalline or microcrystalline and can occur variously aggregated in coal. Pyrite in coal can be found either as macroscopic occurrences (which can be detected with the naked eye) or microscopic forms (which can only be observed under an optical microscope). The most common forms of macroscopic pyrite are the so-called "sulfur balls" or nodules, roughly spherical in shape, which can range in size from a few millimeters to some decimeters, lenses with thicknesses ranging from a few millimeters to several centimeters and up to several decimeters in lateral extent, and veins with variable thickness and extent. The most common forms of microscopic pyrite are (i) finely disseminated pyrite grains (Fig. 1) called euhedral when they are well-formed crystals with sharp, easily recognized faces; (ii) clusters of fine-grained pyrite, called "framboids" (Fig. 2) from the French word *framboise* for raspberry; and

Table 1 Iron sulfides occurring in coals

Name	Composition	Fe (% by mass)	Fe-to-S ratio	Crystal system	Properties
Mackinawite	FeS	63.53	1.742	Tetragonal	Metastable material
Cubic FeS	FeS	63.53	1.742	Cubic	Highly unstable phase precursor
Troilite	FeS	63.53	1.742	Hexagonal	Stoichiometric end member of the $Fe_{1-x}S$ group
Pyrrhotite	$Fe_{1-x}S$ approximately Fe_7S_8 $x > 0.2$	60.38	1.524	Monoclinic	Nonstoichiometric stable group, approxinately F_7S_8
Pyrrhotite	$Fe_{1-x}S$ approximately $Fe_{10}S_{11}$ $x > 0.2$	61.29	1.583	Hexagonal	Nonstoichiometric stable group, approxinately $F_{10}S_{11}$
Smythite	Fe_9S_{11}	58.76	1.425	Hexagonal	Metastable phase related to the $Fe_{1-x}S$ group
Greigite	Fe_3S_4	56.64	1.306	Cubic	Metastable $Fe^{III}Fe^{II}$sulfide; the thiospinel of iron
Pyrite	FeS_2	46.55	0.871	Cubic	Stable iron (II) disulfide known as "fool's gold"
Marcasite	FeS_2	46.55	0.871	Orthorhombic	Metastable iron (II) disulfide

Adapted with permission from Rickard and Luther [59], Copyright 2007 American Chemical Society

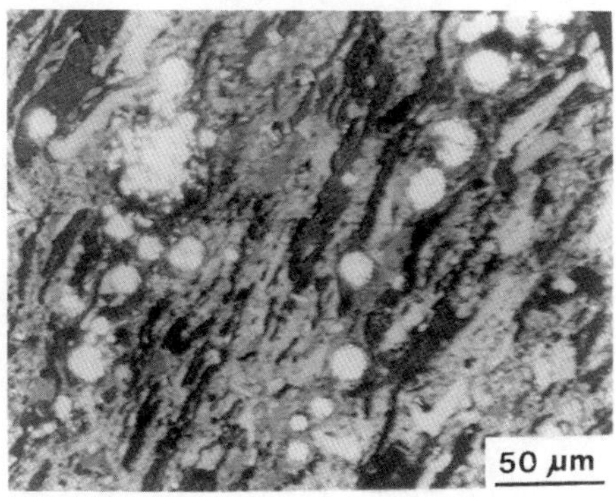

Fig. 1 Micrograph of pyrite grains (*white areas*) in coal matrix. Reflected light, oil immersion

(iii) veinlets (Fig. 3) a few micrometers thick with length in the micrometers range. Excellent descriptions can be found in the literature [24, 52, 71, 80].

The organic sulfur is combined with the coal substance. Organic sulfur atoms may be covalently bound to the atoms of the carbon matrix as thiols, sulfides, disulfides, thiopyrones ([21]; cited by [11]), and complex thiophenic ring systems; several coals contain aromatic heterocyclic compounds with the C–S bond; typical of these compounds seems to be dibenzothiophene (DBT [33, 46]). Thioether bridges [81] have also been shown to exist. According to Ghosh and Prelas [22]

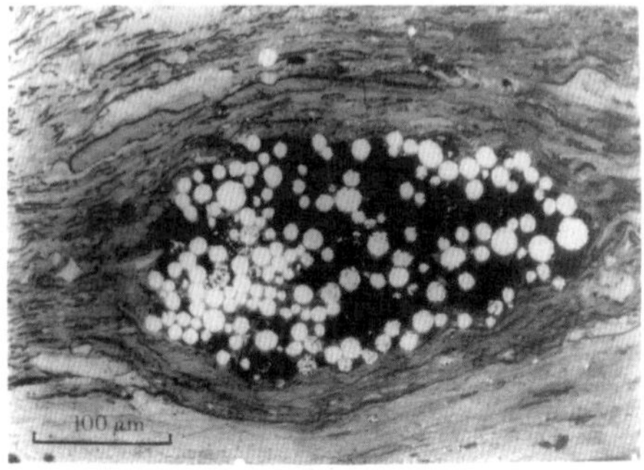

Fig. 2 Micrograph of framboidal pyrite (*white areas*) in coal matrix. Reflected light, oil immersion

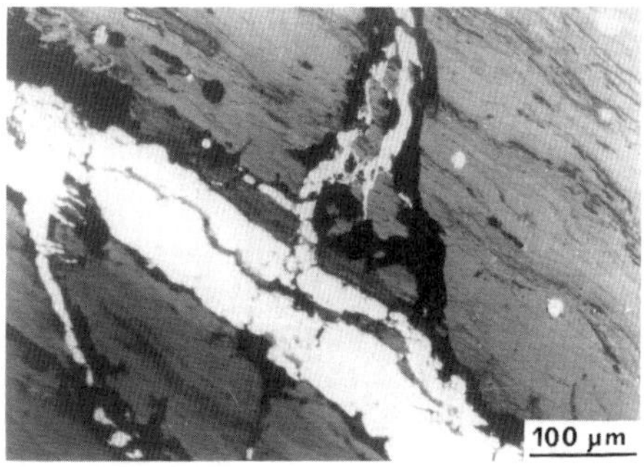

Fig. 3 Micrograph of pyrite veinlet (*white areas*) in coal matrix. Reflected light, oil immersion

sulfur in coals ranges from less than 1 % (low sulfur coals) to about 7 % (high sulfur coals) and organic sulfur ranges from 30 to 70 % of the total sulfur [5]. Figure 4 (modified and redrawn from [70]) is an idealized picture of a coal "macromolecule" that simply gives an idea of how sulfur can be physically and chemically present in coal.

1.1 The Removal of Pyritic Sulfur

Pyritic sulfur is chemically independent of the coal matrix as it is simply disseminated within it and thus physically separated from the coal macerals.

Pyritic sulfur can be removed by conventional mineral processing methods, usually gravity separation or flotation, although some researchers claimed to have successfully applied magnetic separation [47]. The condition for the successful application of mineral dressing processes is the complete liberation of the pyrite grains. This sets a limit on the mesh-of-grind required, as any middlings represent a drawback in the sense that their inclusion in the product reduces its commercial value whereas their rejection decreases coal recovery.

For the benefit of readers who are not familiar with the technical expressions of minerals beneficiation, mesh-of-grind is defined as "the optimum particle size

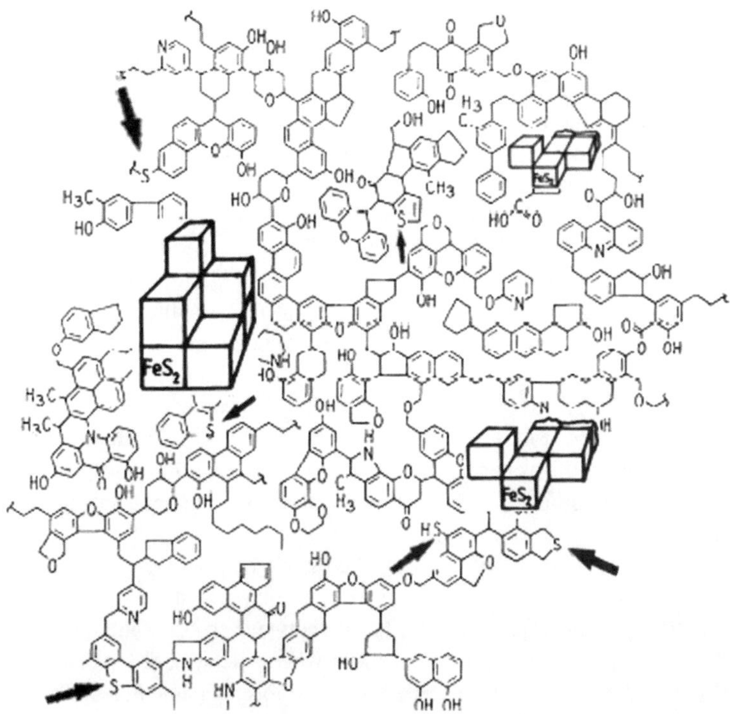

Fig. 4 Shinn's model of coal. The distances are not in scale. The *drawing* is indicative only of the existence of the pyrite inclusions. *Black arrows* point to the covalently bound sulfur atoms (Modified and redrawn from Shinn [70], Copyright 1984 with permission from Elsevier)

resulting from a specific grinding operation, stated in terms of percent of material passing (or alternatively being retained on) a given size screen. The mesh-of-grind is the liberation mesh decided as correct for commercial treatment of the material [1].

In addition, it should be noted that the technical difficulty and process costs of gravity separation and even flotation increase with decreasing mesh-of-grind. Lower limits can be considered as 1 mm for gravity separation and 74 μm for flotation.

In the 1940s and early 1950s it was discovered that a number of microbial strains [16, 44, 75] were able to enhance the kinetics of metal sulfides, including pyrite solubilization in water (for review, [61]). One of those strains that proved to be particularly effective is the acidophilic, mesophilic, and chemolithoautotrophic *Acidithiobacillus ferrooxidans*. In a simplified (somewhat improper) way, it can be said that these micro-organisms act as biological catalysts of pyrite oxidation and solubilization. The biologically catalyzed oxidation of pyrite can be described by the overall reaction

$$2FeS_2 + 7.5O_2 + H_2O \rightarrow 2Fe^{3+} + 4SO_4^{2-} + 2H^+$$

Evidence was provided of the net production of one H^+ per mol of FeS_2 oxidized and of the end products Fe $^{3+}$ and $2SO_4^{-2}$. Details on this process can be found by the interested reader in Chapter "Biomining". Overviews of the bacteria from a number of different taxonomic groups, namely, the genera of *Thiobacillus* (*Acidithiobacillus*), according to the recent new classification proposed by Kelly and Wood [37] and *Leptospirillum* and archaea, several of which are involved in the removal of inorganic and organic sulfur compounds from coal, can be found in Karavaiko and Lobyreva [33], Schippers [67], Johnson and Hallberg [31], and Hedrich et al. [27]. Mixed cultures warrant investigation as they may prove somewhat beneficial, as indicated for the case of metal sulfide bioleaching [53].

The resort to microbiological mediation overcomes the drawbacks of conventional mineral processing methods mentioned above, as the exposure of just part of the pyrite grains' surfaces is clearly sufficient to grant access of the microbial cells thereto, thus ensuring solubilization of the whole pyrite grain, reducing the need for fine grinding. In this regard coal porosity plays an important role: the exposure of part of the pyrite grains at the pores' edges (Fig. 5 [28]) being sufficient for their biosolubilization.

Also worthy of note is the fact that the action of micro-organisms is favored whenever they are able to excrete the extracellular polymeric substances (EPS) (Chapter "Biomining") that acts as a bridge between the cell and the pyrite grain [60]. This should also be taken into account when selecting the type of bioreactor to be used, as described later. The EPS is possibly mechanically not very strong. The author is not aware of any "ad hoc" publications on this subject, however, a paper [45] on a similar subject could provide some indications. When economically convenient and the degree of intergrowth of pyrite with coal is on the order of the centimeter, comminution can be reduced and bioleaching can be performed in coal piles or heaps [8].

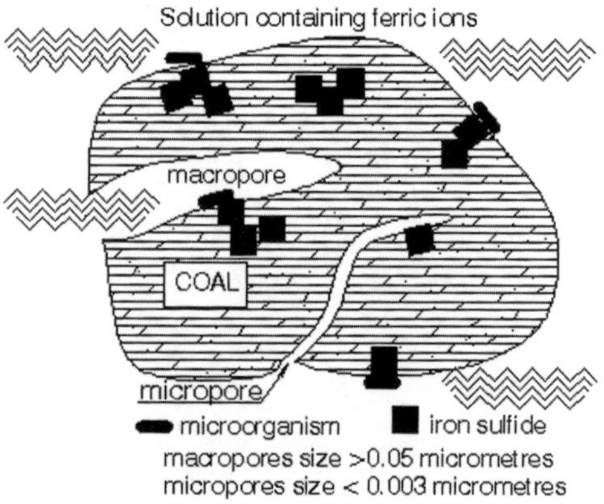

Fig. 5 Schematic of porosity in coal. Only attached bacteria are shown, but active swimming also occurs. Most pyrite particles may be much larger than the bacterial cells. Drawing not in scale (Modified and redrawn from [28] Copyright 1987 with permission from VCH Verlagsgesellschaft mbH)

1.2 The Removal of Organic Sulfur

Organic sulfur atoms are covalently bound to the atoms of the carbon matrix as thiols, sulfides, disulfides, and complex thiophenic ring systems. Several coals contain aromatic heterocyclic compounds with the C–S bond; typical of these is dibenzothiophene [33, 46]. The removal of organic sulfur therefore requires the preliminary breakage of those C–S bonds. Let us first recall, for the benefit of those readers who are not biochemists, that metabolic pathways are a series of chemical reactions, occurring within a microbial cell, mediated by enzymes. In pyrite solubilization the micro-organism gets the energy needed for its metabolism mainly from the oxidation of pyritic iron in the divalent state. Hence, its metabolic pathway had already been identified in early bioleaching times. For the organic sulfur compounds identifying the metabolic pathway is more complex. In effect, depending on the compound in which sulfur is chemically bound, the pathway can be quite different. For this reason, model compounds have been, and continue to be, investigated.

As far as the author is aware, no pilot testing of organic sulfur removal from coal has been undertaken thus far. Therefore an overview of the most significant research conducted to date and the results obtained are considered useful, chiefly because some of them appear to be somewhat controversial.

The origins of research on organic sulfur removal by means of micro-organisms date back to the 1950s, and concerned investigations aimed at removing organic sulfur from petroleum. Research on the compounds contained in crude oil produced evidence of the presence of dibenzothiophene and this compound was selected as a model compound for laboratory investigations into the possibilities of C–S bond disruption via microbial attack, and the formation of water-soluble compounds.

A number of micro-organisms have been claimed over the years to be capable of breaking the C–S bonds in coal: *Beijerinckia* [43], *Pseudomonas sp.* [30, 32, 41, 51], *Acinetobacter* [29], *Desulfovibrio desulfuricans*, *Rhodococcus rhodochrous* [25], *Sulfolobus* [34, 35], *Brevibacterium, Cunninghamella elegans* [17], *Escherichia coli*, *Rhizobium* [20].

The papers published by Isbister [29], Isbister and Doyle [30], and by Kilbane [38] raised a great deal of interest at the time, as they claimed to have obtained very encouraging results working with mutant strains on DBT and also on coal. However, their findings were not subsequently confirmed.

In the late 1990s a comprehensive collaborative coal desulfurization project (Jouf 0039, *Microbial Desulfurization of Coal*) was launched, funded by the European Commission which included as partners Germany's Deutsche Montan Technologie (D.M.T., Essen), Italy (Geoengineering Department of the University of Cagliari, and the Italian Agency for Hydrocarbons, Ente Nazionale Idrocarburi (E.N.I.), The Netherlands (University of Delft), and the United Kingdom (Stevenage Research Laboratories). Within the framework of this project, the possibilities of organic

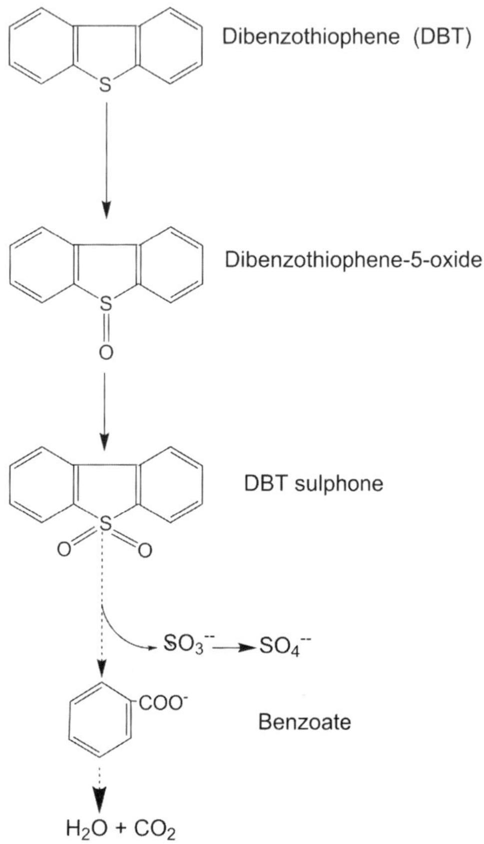

Fig. 6 Metabolic pathway of DBT degradation by *Brevibacterium* sp. DO according to Van Afferden (Modified and redrawn from Van Afferden et al. [79], Copyright 1990 with permission from Springer)

sulfur removal were thoroughly investigated. In a series of papers [77–79] the researchers from the German team produced evidence of having isolated a mixed culture, named FODO, consisting of an *Alcaligenes denitrificans* subspecies and a *Brevibacterium* species capable of utilizing dibenzothiophene as the sole sulfur source for growth, and benzoate was used as the carbon source, and a pure *Brevibacterium* sp. culture able to use dibenzothiophene as the sole source of carbon, sulfur, and energy for growth. The remarkable feature of this work was that for the first time evidence was provided of a sulfur-specific attack on DBT by a two-species bacterial community that utilizes DBT as the sole source of sulfur. The proposed pathway—developed after the metabolites of dibenzothiophene degradation were identified as dibenzothiophene-5-oxide, dibenzothiophene-5-dioxide, and benzoate by co-chromatography, UV spectroscopy, and gas chromatography mass spectrometry analyses—is shown in Fig. 6. Table 2 summarizes some of the most significant results of organic sulfur bioremoval and points out that the best results were obtained when *Brevibacterium* sp. or *Pseudomonas* sp. were used.

Table 2 Summary for organic sulfur bioremoval

Organism	Organic sulfur removal (%)	Substrate for adaptation/ enrichment	Reference
Pseudomonas janji DDC279	>95	DBT	[41]
Bacterial mixed culture	30	DBT	[13]
Acidithiobacillus ferrooxidans	56	–	[23]
Pseudomonas sp.	47	DBT	[29]
Sulfolobus acidocaldarius	10	DBT	[36]
Defined bacterial species	<7	Thiophene, cysteine, benzene, sulfonic acid	[40]
Bacillus sp.	36	DBT	[12]
Pseudomonas sp.			
Micrococcus sp.			
Pseudomonas putida	37	DBT	[58]
Hansenula sp.	<46	Cysteine, methione	[72]
Cryptococcus albidus		Thiophene, DBT	
Gram-negative bacteria	0	DBT	[73]
Bacterial mixed culture	33	DBT	[66]
Brevibacterium sp. (named "DO")	>95	DBT and thiamine	[79]
Fungus	<20	Not specified	

Modified from Klein et al. [39], Copyright 1994 with permission from Elsevier

For DBT degradation the metabolic pathway 4S shown in Fig. 7 was proposed by Kodama et al. [42]. Another pathway, called the 2S pathway (Fig. 8), termed the "oxidative" pathway, is a carbon-targeted reaction and, as such, has little relevance for coal desulfurization technologies [82].

As far as the 4S pathway is concerned, Andrews and Datta [3] presented an analysis of the choice mechanism for sulfur removal from DBT and showed that the free energy values of the intermediaries of the 4S pathway as calculated by the chemical thermodynamics methods, imply that this process is generally thermodynamically favorable with the exception of the step from DBT-5-oxide to sulfone. This conversion requires about 100 kJ/mole and, as the amount of energy involved is fairly large, the reaction does not occur spontaneously, unless some external agent takes part in the process. Therefore, according to these authors it only can be said that DBT as a model compound can lead to misleading conclusions.

Organic sulfur bioremoval was also tackled, very likely using the shaken flasks technique, using unspecified fungi [18] with results that the authors claimed to be better than those obtained with *Sulfolobus* but on which more information would be desirable.

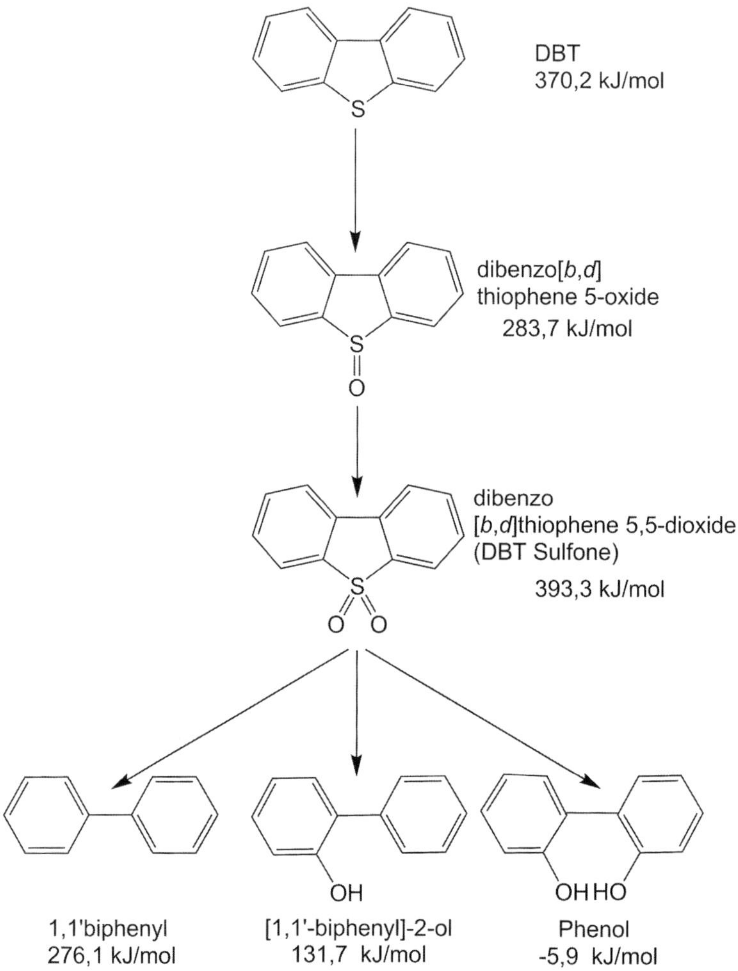

Fig. 7 Metabolic pathway "4S" and its thermodynamic parameters (Modified and redrawn from Andrews and Datta [3], Copyright 1991 with permission from EPRI)

2 Semi-Commercial Coal Biodepyritization Operation

The encouraging results of around half a century's basic research on coal biodepyritization justified the move to continuous testing, initially at the laboratory scale and subsequently at the semi-commercial pilot scale. The first laboratory-scale continuous biodepyritization plant was designed and operated in the late 1980s at Deutsche Montan Technologie (DMT) in Germany [6, 76]. The predominant micro-organism in the mixed culture employed was *A. ferrooxidans*: the equipment consisted of a cascade of eight 20-dm^3 Pachuca-type units which achieved pyrite conversions of up to 70 %. The authors claim that, at a slurry density of

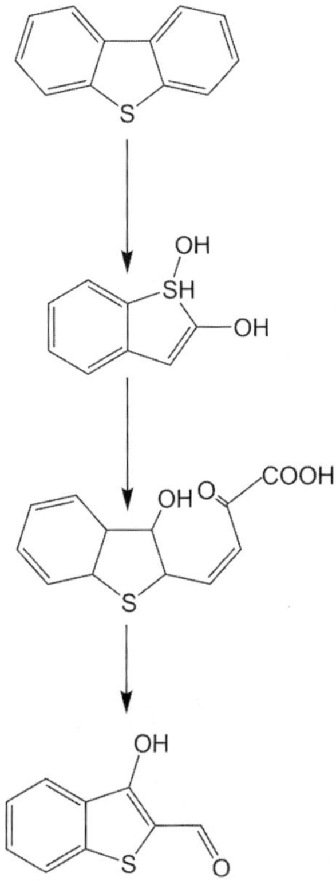

Fig. 8 Metabolic pathway "2S" (Modified and redrawn from Kilbane [38], Copyright 1990 with permission from Elsevier)

20 % (w/w), about 700 mg of ferrous iron were solubilized in 80 h. The Pachuca reactors were originally developed and commercially applied by the hydrometallurgists; their name derives from the Mexican city of *Pachuca,* where they were first used for precious metals leaching. A Pachuca tank is a cylindrical tank with a conical bottom. It contains a pipe that is coaxial with the leaching tank and open at both ends; compressed air is introduced at the lower end of this pipe, which behaves as an air lift. The density of the pulp within the pipe is less than that of the pulp surrounding it because the column of air bubbles contained in the pipe, and the pressure of denser pulp, causes the pulp in the central pipe to rise and overflow, thus circulating the entire charge" [1].

They are substantially air-lift reactors (Fig. 9a) and were thoroughly investigated ([65, 69]; [68]). Figure 9b shows a diagram of the Pyrex glass Pachuca bioreactors designed and constructed by DMT.

Almost contemporaneously, a technoeconomic analysis of the continuous biodepyritization process was published by the research team at the University of Delft [7] and a proposal for scale-up of reactors for coal depyritization was also published [2]. To be precise, it should be mentioned that a bench-scale

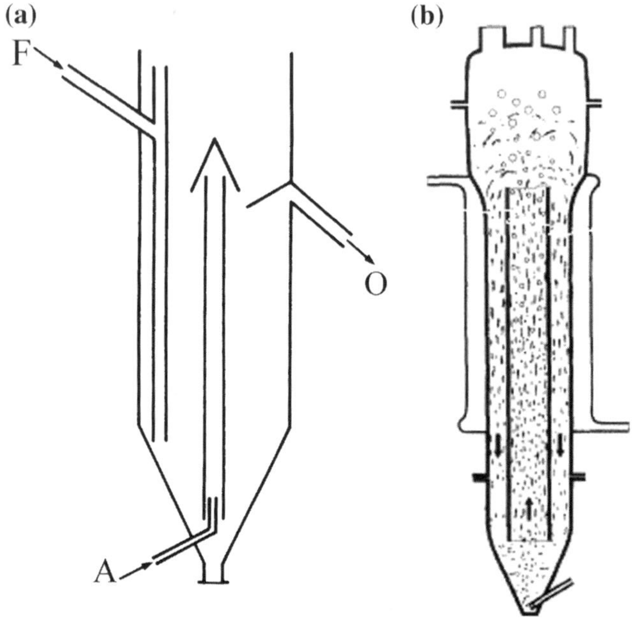

Fig. 9 **a** Pachuca-type bioreactor: principle: *A* air injector, *F* Feed inlet, *O* depyritized coal suspension outlet. **b** Pachuca-type bioreactor: DMT model (Modified from Beyer et al. [6], Copyright 1986 with permission from Springer)

depyritization test was recently carried out [56] confirming the feasibility of the process but with less attractive results than those obtained by the DMT. More recently, Cardona and Marquez [9] applied bioleaching to coal depyritization, although operating on 10 % solids suspensions. Interestingly, these workers used a consortium of native micro-organisms.

The first semi-commercial continuous biodepyritization operation was designed at the end of the 1980s by the partners in the EU Project JOUF-0039 (Microbial Desulfurization of Coal). The plant, erected in an area of the chemical complex of the Italian Agency EniChem, on the outskirts of the town of Porto Torres in Northern Sardinia, had a capacity of 50 kg raw coal per hour and consisted of three main sections: a comminution bay, a bioreactor bay, and a reject water purification and disposal system. These are described in the following (Figs. 10 and 11).

2.1 The Comminution Bay

The comminution bay consisted of a 4,000-kilo head bin wherefrom the raw coal, maximum size 100 mm, was conveyed by a belt feeder at a mass flow rate regulated by an automatic scale.

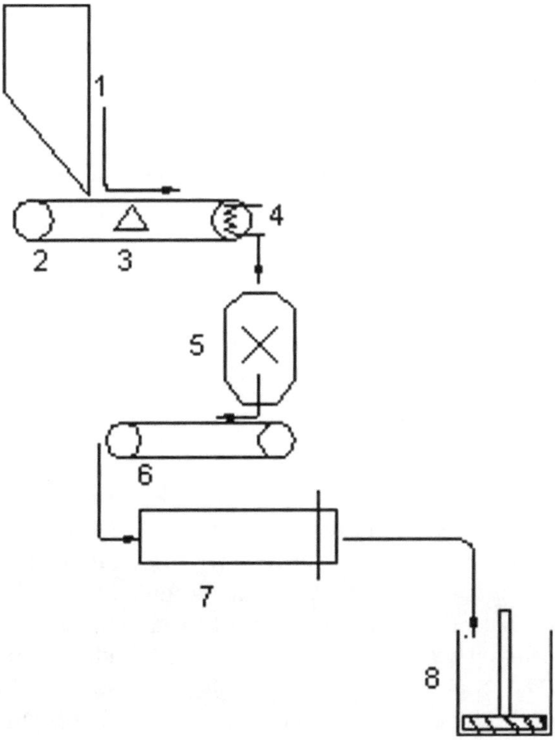

Fig. 10 Porto Torres biodepyritization pilot plant: Coal comminution bay. *1* head bin, *2* belt feeder, *3* automatic scale, *4* magnetic separator, *5* hammer mill, *6* belt conveyor, *7* wet ball mill, *8* preparation tank (Modified and redrawn from Rossi [62], and Loi et al. [49], Copyright 1993 with permission from Elsevier)

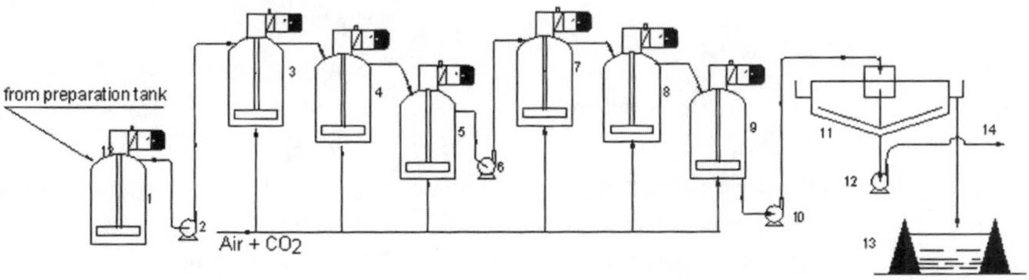

Fig. 11 Porto Torres biodepyritization pilot plant: bioreactor and coal dewatering bay. *1* propagator, *2* pump, *3–9* stirred tank bioreactors, *10* pump, *11* rake thickener, *12* diaphragm pump, *13* settling pond, *14* to stock-pile (Modified and redrawn from Rossi [62], and Loi et al. [49], Copyright 1993 with permission from Elsevier)

After removal of any tramp iron by means of a magnetic separator, the coal was crushed in a hammer mill to a top size of 4 mm. Further size reduction to the desired mesh-of-grind was performed in a wet ball mill, fed by a conveyor with the

Table 3 Characteristics of Sulcis coal

Parameter	Percent
Total moisture	7.33
Volatile content	42.95
Ash	11.44
Fixed carbon	45.61
Total carbon	61.70
Hydrogen	4.54
Nitrogen	1.15
Oxygen	14.53
Total sulfur	6.64
Sulfate sulfur	0.27
Pyritic sulfur	1.74
Organic sulfur	4.63

ground product of the hammer mill. The coal (a batch of 200 tonnes) used for the first run, designed to last a whole year, was supplied by the Seruci coal mine in the Sulcis subbituminous coal basin located in southern Sardinia [10]. Part of the pyrite contained in this coal is very finely intergrown with the matrix and pyrite exposure requires grinding to 100 % passing 40 μm. Table 3 shows the typical composition of this coal. The ground product flowed from the ball mill by gravity to a 4 m^3 preparation tank where the solids concentration was adjusted to the desired value. The required nutrients and sulfuric acid solution for adjusting and maintaining the pH of the suspension in the 2.10–2.35 range, optimum acidity for *Acidithiobacillus ferrooxidans*, were then added.

2.2 The Bioreactor Bay, Depyritized Coal Stockpile, and Reject Water Disposal

The thus-prepared suspension was pumped to a 7.5-m^3 mixing tank, a "propagator," and inoculated with a microbial strain of *A. ferrooxidans* (very likely a community also containing some *Leptospirillum ferrooxidans* [61]) supplied by the biohydrometallurgy laboratory of the University of Cagliari, and from there conveyed to the bioreactors. Pyrite biosolubilization was carried out in six 7.5-m^3 stirred tank bioreactors, 4 m high and 2 m in diameter, arranged in cascade. Each bioreactor was provided with a 1-m diameter six-bladed Rushton-type impeller [19] driven by an electric motor coupled to a speed variator. The cylindrical bioreactor tanks were equipped with four baffles at 90° from one another and water jackets, where water at the desired temperature could be circulated in case of need.

Impeller speed and the instantaneous power consumption of each motor were monitored on the control panel and recorded. More details on these reactors and on the energy consumption at the various solids concentrations of the coal suspensions can be found in Orsi et al. [55] and Loi et al. [48, 49].

As anticipated by other researchers [2, 6], solids concentration is a critical technical and economic parameter in bioleaching in general [26, 54, 57, 62, 63] and especially in coal biodepyritization [4, 8, 48, 49]. In fact, owing to the relatively low economic value of coal, high solids concentrations need to be pursued, resulting in smaller machine sizes and consequently lower headroom requirements and investment costs. However, it seemed that, with current bioreactors, 20 % solids was an insurmountable limit. It was therefore considered expedient to thoroughly investigate this parameter, and runs lasting at least 15 days each were carried out at 6.5, 13.5, 14.6, 19, 29.5, and 41.5 solids concentration.

Pyrite removal increased with the number of bioreactors in the cascade and achieved 90 %, for all runs, in the fifth bioreactor, any increase in the next bioreactor being negligibly small. Most probably, the 10 % pyrite that remained in the coal consisted of pyrite grains too minute to be liberated at the mesh-of-grind of the ball mill; observations under the optical microscope produced evidence of the existence of this kind of middlings. It was therefore considered quite reasonable to carry out all calculations on the basis of a cascade comprising five bioreactors. For a suspension flow rate of 250 L h^{-1}, the average residence time was calculated to be 8.25 days and the iron pyrite solubilization rate 36 mg dm^{-3} h^{-1}, corresponding to a pyrite solubilization rate constant of 1.53×10^{-2} h^{-1}. The power requirement per bioreactor operating with a 40 % solids pulp was 4.0 kW, with $\cos \varphi = 0.76$ in alternating currents ($\cos \varphi$ is the power factor, where φ is the phase difference between electromotive force and current); in electric power contracts a clause frequently sets forth that if a customer permits the average power factor of the load used to fall below a specified value, a penalty charge will be made [1]. Hence, the power required for depyritizing 100 kg h^{-1} coal at 40 % solids in the bioreactor section was calculated as $4.0 \times 5 = 20$ kWh, that is, 200 kWh per tonne dry coal. This is already a very encouraging economic result. However, at the present cost of electricity the cost of the power requirement is still too high, at least in some countries. More information on plant performance and on the calculations can be found elsewhere [48, 49].

The depyritized coal suspension was finally pumped to a thickener. As the pH of the liquid phase of the suspension was below 1.7, it was necessary to dispose of the latter in a settling pond for separation of the clarified liquid phase. In effect, the solid phase does not need to be completely dessicated, one of its most convenient uses being as coal–water mixtures; however, owing to the high acidity of the liquid phase—which can obviously be harmful to the machinery—the percentage of the latter should be kept under strict control. A diaphragm pump was used to pump the thickened solids out of the thickener and convey them to a stockpile.

It should also be observed that the processed coal contains less ash than the feed, but it appears to have undergone considerable oxidation (Table 4).

In the biohydrometallurgy laboratory, the problem of the bioreactor's headroom and power requirement has been addressed and a novel type of bioreactor, named "Biorotor" [50, 64] was designed, developed, and tested. This bioreactor, consisting of a cascade of rotating drums, is characterized by very gentle stirring with minimal shear stresses within the suspension hence with very low stresses on the

Table 4 Comparison of the biodepyritization product and the feed sample

Coal characteristic features	Coal feed sample	Depyritized coal	Variation (%)
Pyritic sulfur (%)	2.36	0.23	−90.25
Ash (%)	15.28	9.31	−39.07
Fixed carbon (%)	40.82	47.51	+16.96
Upper calorific value (kJ kg^{-1})	25.937	26.041	+0.40

Modified and redrawn from Loi et al. [48, 49], Copyright 1994, with permission from Elsevier

EPS layers. Two preliminary tests, carried out with 40 % solids suspensions, one consisting of a mineral sulfides feed, and the other of the same Sulcis coal used in the pilot plant, proved to be very encouraging, yielding the same percent sulfide removal and lower power requirements.

The development of a bioreactor suitable for depyritizing coal suspensions with solids concentrations higher than 40 % and lower power requirements and very moderate stirring obviously warrants attention.

3 Concluding Remarks and Outlook

The feasibility of pyritic sulfur bioremoval has been ascertained at the laboratory scale and confirmed by testing in a semi-commercial pilot plant on a somewhat difficult to process subbituminous coal from the Sulcis basin (Sardinia, Italy). The results of the pilot plant operation described above are very encouraging, although the costs of power per tonne of coal processed appear to be high, at least at current electricity prices. Bioreactors tailor-made for the biodepyritization process would likely require less power than the conventional stirred tank reactors installed in the plant. Evidence has been provided, in several decades of research conducted worldwide, that mesophiles are the most suitable micro-organisms, although process kinetics can be enhanced at high temperatures using thermophiles [14, 15, 34, 35]. In actual fact, as temperature is increased, so the detrimental effects of jarosite precipitation emerge as solubilization kinetics slow down. One area that has received little attention up to now, but that warrants careful investigation, is microbial consortia.

Research on bioremoval of organic sulfur has not enjoyed the same success, notwithstanding the major efforts undertaken in all industrial countries also on account of its being very closely related with the problem of organic sulfur removal from liquid fossil fuels. Most past and ongoing research is carried out in the laboratory and concerns model compounds, especially DBT. The choice of DBT has received some criticism on the grounds of chemical thermodynamics calculations referred to the 4S metabolic pathway proposed as the mechanism of choice for removing sulfur from DBT. Thus it is recommended that investigations on other model compounds and on bacterial consortia should be completed.

At the end of the Alghero Symposium, a "Strategic Document" was drawn up by a team of experts in compliance with the request by EC executives who had supported the EC sponsorship. This document is an interesting outlook for future research as one can infer from the following excerpt reproducing its introduction. As an exercise during the Fourth International Symposium on the Biological Processing of Fossil Fuels, the Organizing and Executive Committee attempted to generate a document that would assist investigators in identifying research areas which are market-driven and have a higher probability of receiving support. This exercise was deemed of high importance since there has been a drastic decline in funding of the classical research that has been represented by this symposium. Members of the International Scientific Committee, Organizing and Executive Committee, and symposium participants collaboratively discussed a variety of strategic issues related to the biological processing of fossil fuels. These discussions focussed on four key issues:

1. state-of-the-art of biological-based processing,
2. technical and economic bottlenecks to the successful commercialization of biological-based processes for fossil fuels,
3. R&D programs which could overcome such bottlenecks, and
4. market forces which drive the development of biological-based processes. Each of these issues were considered in the context of four major research areas:

    1. gas processing, including metals, SO_2, NOx, syngas, CO_2, H_2S and HCI
    2. biodesulphurization processes, including coal and oil
    3. metal-related processes, including coal, ash and oil, and
    4. biosolubilization, bioliquefaction and biogasification.

Information was extracted from the six technical sessions and two workshops in order to synthesize a synopsis of the relevant issues [74].

References

1. Am Geol Inst (1997) Dictionary of mining, mineral and related terms, 2nd edn. The American Geological Institute in cooperation with the Society for Mining, Metallurgy and Exploration
2. Andrews G (1989) Reactors for scaling up coal bioprocessing. EPRI symposium on biological processing of coal and coal-derived substances, Palo Alto, CA, USA, pp 1–17
3. Andrews G, Datta R (1991) Mass and energy balances for coal-based microbial systems. In: Proceedings of the second international symposium on biological processing of coal, San Diego, CA, USA. EPRI US Department of Energy: 2-1-2-21
4. Andrews G, Quintana J (1990) Mixing and mass transfer in an aerated trough bioreactor. In: Proceedings of the first international symposium on biological processing of coal. Orlando, FL, USA; Palo Alto, CA, USA. Electric Power Research Institute: 5-69-5-85
5. ASTM International D2492-02 (2007) Standard Test Method for Forms of Sulfur in Coal. West Conshohocken, PA 19428-2959
6. Beyer M, Ebner HG, Klein J (1986) Influence of pulp density and bioreactor design on microbial desulphurization of coal. Appl Microbiol Biotechnol 24:342–346

7. Bos P, Huber TF, Kos CH, Ras C, Kuenen JG (1986) A Dutch feasibility study on microbial coal desulphurization. In: Lawrence RD, Branion RMR, Ebner HGR, Ritcey GM (eds) Process metallurgy. Elsevier, Amsterdam
8. Cara J, Carballo MT, Mora A, Bonilla D, Escolano O, Garcia Frutos FJ (2005) Biodesulphurisation of high sulphur coal by heap leaching. Fuel 84:1905–1910
9. Cardona IC, Marquez MA (2009) Biodesulfurization of two Colombian coals with native microorganisms. Fuel Procesa Technol 90:1099–1106
10. Carta M, Del Fa' C, Agus M, Carbini P (1978) Studio di una campionatura del I fascio della Miniera di Seruci del Bacino carbonifero del Sulcis. Resoconti dell'Associazione Mineraria Sarda LXXXIV(1):111–169
11. Casagrande DJ (1987) Sulphur in peat and coal. In: Scott AC (ed) Coal and coal-bearing Strata: Recent Advances. Geological Society, London
12. Chandra D, Mishra AK (1988) Desulfurization of coal by bacterial means. Resour Conserv Recycl 1:293–308
13. Chandra D, Roy P, Mishra AK, Chakrabarti JN, Sengupta B (1979) Microbial removal of organic sulfur from coal. Fuel 58:549–550
14. Chen CCY, Skidmore DR (1989) Sulfur leaching by thermophilic microbes of coal particles varying in size. In: Schreiner BJ, Doyle FM, Kawatra SK (eds) Biotechnology in minerals and metal processing. Society of Mining Engineers Inc., Littleton
15. Chen CCY, Skidmore DR (1990) Microbial coal desulfurization with thermophilic microorgnisms. In: Wise DL (ed) Bioprocess biotreat coal. Marcel Dekker Inc, New York
16. Colmer AR, Hinkle ME (1947) The role of micrioorganisms in acid mine drainage. A preliminary Report. Science 106(2751):253–256
17. Cutright TJ (1995) Polycyclic aromatic hydrocarbon biodegradation and kinetics using *Cunninghamella echinukata var, elegans*. Int Biodeterior Biodegrad 95:397–408
18. De-wen H, Yan-jie L, Li-yuan C, Yun-yan W, Yan J, Bing P (2006) Desulfurization of coal by fungus. In: Kongolim F, Reddy RG (eds) Sohn international symposium advanced processing of metals and materials, thermo physicochem princ, vol 2. TMS, Warrendale
19 Edwards MF, Baker MR (1992) A review of liquid mixing equipment. In: Harnby N, Edwards MF, Nienow AW (eds) Mixing in the process industries, Butterworths Heinemann, London
20. Frassinetti S, Setti L, Corti A, Farrinelli P, Montevecchi P, Vallini G (1998) Biodegradation of dibenzothiophene by a nodulating isolate of *Rhizobium meliloti*. Can J Microbiol 44(3):289–297
21. Given PH, Wyss WF (1961) The chemistry of sulfur in coal. British Coal Util Res Assoc Mon Bull 25:165–179.
22. Ghosh TK, Prelas MA (2009) Coal: Energy, resources and systems: volume 1: Fundamentals and non-renewable resources. Springer, Hamburg
23. Göckay CF, Yurteri RN (1983) Microbial desulfurization of lugnites by a thermophilic bacterium. FUEL 62:1223–1224
24. Gray RJ, Schapiro N, Dole Coe G (1963) Distribution and forms of sulfur in high volatile Pittsburgh seam coal. Trans AIME 226:113–121
25. Haghighat FH, Fereshteh E, Mahnaz M (2003) Isolation of a dibenzothiophene desulfurizing bacterium from soil of Tabriz oil refinery. Iranian J Biotechnol 1(2):121–134
26. Hansford GS, Bailey AD (1993) Oxygen transfer limitation of bio-oxidation at high solids concentration. Torma AE, Wey JE, Lakshmanan VI (eds) Proceedings of the international biohydrometal symposium. The Metals, Minerals and Materials Society, Warrendale
27. Hedrich S, Schlömann M, Johnson DB (2011) The iron-oxidizing proteobacteria. Microbiol 157(Pt 6):1551–1564
28. Höne H-J, Beyer M, Ebner HG, Klein J, Jüntgen H (1957) Microbial deaulphurization of coal—development and application of a slurry reactor. Chem Ing Technol 10:173–179
29. Isbister JD (1985) Mutant microorganism and its use in removing organic sulfur compounds. Atlantic Research Corporation. USA Patent (4808535)

30. Isbister JD, Doyle RC (1989) Mutant microorganism and its use in removing organic sulfur compounds. Atlantic Research Corporation. USA Patent (4562156)
31. Johnson DB, Hallberg KB (2009) Carbon, iron and sulfur metabolism in acidophilic microorganisms. Adv Microb Physiol 54:201–255
32. Julia De Fatima WM, Setti L, Lanzarini G, Pifferi PG (1996) Dibenzothiophene biodegradation by a pseudomonas sp.in poorly degradable organic solvents. Process Biochem 31(7):711–717
33. Karavaiko GI, Lobyreva LB (1994) An overview of the bacteria and archaea involved in removal of inorganic and organic sulfur compounds from coal. Fuel Process Technol 40:167–182
34. Kargi F (1982) Microbiological coal desulphurization. Enzyme Microb Technol 4:13–19
35. Kargi F (1990) Use of *Sulfolobus acidocaldarius*f for microbial desulfurization of coal. In: Wise DL (ed) Bioprocessing and biotreatment of coal., Marcel Dekker, Inc., New York
36 Kargi F, Robinson JM (1986) Removal of organic sulphur from bituminous coal. FUEL 65:397–399
37. Kelly DP, Wood AP (2000) Reclassification of some species of *Thiobacillus* to the newly designated genera *Acidithiobacillus* gen.nov., *Halothiobacillus* gen.nov., and *Thermithiobacillus* gen.nov. Int J Syst Evol Microbiol 50:511–516
38. Kilbane JJ II (1990) Sulfur-specific microbial metabolism of organic compounds. Resour Conserv Recycl 3:69–79
39. Klein J, van Afferden M, Pfeifer F, Schacht S (1994) Microbial desulfurization of coal and oil. Fuel Process Technol 40(297):310
40. Klubek B, Ochman M, Nabe S, Clark D, Alam K, Abdulrashid N (1988) Microbial removal of organic sulphur from coal. Mineral Matters 10:1–3
41. Kodama K (1970) Microbial conversion of petro-sulfur compounds part III. Agr Biol Chem 34(9):1320–1324
42. Kodama K, Umehara K, Shimizu K, Nakatani S, Minoda Y, Yamada K (1973) Identification of microbial products from dibenzothiophene and its proposed oxidation pathway. Agr Biol Chem 37(1):45–50
43. Laborde A, Gibson DT (1977) Metabolism of dibenzothiophene by *Beijerinkia* species. Appl Environ Microbiol 34(6):783–790
44. Leathen WW, Braley SA, McIntyre LD (1953) The role of bacteria in the formation of acid from certain sulfuritic constituents associated with bituminous coal. II. Ferrous iron oxidizing bacteria. Appl Bacteriol 1:65–68
45. Leung KM, Wanger G, Qiuquan G, Gorby Y, Southam G, Woon M, Jun Yang L (2011) Bacterial nanowires: conductive as silicon, soft as polymer. Soft Matter 7:6617–6621
46. Levine DG, Schlosberg RH, Silbernagel BG (1982) Understanding the chemistry and physics of coal structure (A Review). Proc Natl Acad Sci U S A 79:3365–3370
47. Liu YA, Lin CJ (1976) Assessment of sulfur and ash removal from coals by magnetic separation. IEEE trans Magnetics, MAG 12:538–550
48. Loi G, Mura A, Trois P, Rossi G (1994) Bioreactor performance versus solids concentration in coal biodepyritization. Fuel Process Technol 40:251–260
49. Loi G, Mura A, Rossi G, Trois P (1994) The Porto Torres biodepyritization pilot plant: light and shade of one year operation. Fuel Process Technol 40:261–268
50. Loi G, Rossi A, Trois P, Rossi G (2006) Continuous revolving barrel bioreactor tailored to the bioleaching microorganisms. Miner Metall Process 23(4):196–202
51. Monticello DJ, Bakker D, Finnerty WR (1985) Plasmid mediated degradation of dibenzothiophene by *Pseudomonas* species. Appl Environ Microbiol 49(4):756–760
52. Neavel R (1966) Sulfur in coal: its distribution in the seam and in mine products. Dissertation, Pennsylvania State University
53. Okibe N, Gericke M, Hallberg, KB, Barrie Johnson, D (2003) Enumeration and characterization of acidophilic microorganisms isolated from a pilot plant stirred-tank bioleaching operation. Appl Environ Microbiol 66(4):1936–1943

54. Oolman T (1993) Bioreactor design and scale up. Applications in minerals bioleaching. Torma AE, Wey JE, Lakshmanan VI (eds) Proceedings of the international symposium on biohydrometallurgy, vol 1. The Metals, Minerals and Materials Society, Warrendale, pp 401–415
55. Orsi N, Rossi G, Trois P, Va lenti PD, Zecchin A (1991) Coal biodesulphurization: design criteria of a pilot plant. Res Conserv Recycl 5:211–230
56. Pandey RA, Raman VK, Bodkhe SY, Handa BK, Bal AS (2005) Microbial desulfurization of coal containing pyritic sulfur in a continuously operated bench scale coal slurry reactor. Fuel 84:81–87
57. Pinches A, Huberts L, van Staden M, Muehlbauer RM (1991) Development and optimization of bacterial oxidation processes using small-scale pilot plants. Colloquium bacterial oxidation. The South African Institute of Mining and Metallurgy, Johannesburg, pp 67–89
58. Rai C, Reyniers JP (1988) Microbial desulfurization of coal by organisms of the genus Pseudomonas. Biotechnol Progress 4:225–230
59. Rickard DT, Luther GW III (2007) Chemistry of Iron Sulfides. Chem Rev 107:514–562
60. Rohwerder T, Sand W (2007) Mechanisms and biochemical fundamentals of bacterial metal sulfide oxidation. In: Donati E, Sand W, (eds) Microbial processing of metal sulfides. Springer, Dordrecht, pp 35–58 (The Netherlands)
61. Rossi G (1990) Biohydrometallurgy. McGraw-Hill Book Co., Hamburg
62. Rossi G (1993) Biodepyritization of coal: achievements and problems. Fuel 72(12):1581–1592
63. Rossi G (1993) La miscelazione nei reattori per bioidrometallurgia. ICP—Riv Ind Chimi 20(2):76–86
64. Rossi G (2001) The design of bioreactors. Hydrometall 59:217–231
65. Roy GG, Sheckhar R (1996) Oxygen mass transfer in air-agitated Pachuca tanks—Part 1: Laboratory-scale experimental measurements. Part 2: Mathematical modelling of mass-transfer coefficients. Trans Inst Min Metall 105 (Section C):9–21
66. Runnion KN, Combi JD (1990) Microbial removal of organic sulfur from coal. 2/62–2/76
67. Schippers A (2007) Microorganisms involved in bioleaching and nucleic acid-based molecular methods for their identification and quantification. In: Donati E, Sand W (eds) Microbial processing of metal sulfides. Springer, Dordrecht, pp 3–33
68. Sheckhar R, Evans JW (1989) Fluid flow in pachuca (air agitated)tanks: Part I. Laboratory-scale experimental measurements. Metall-Trans- B, pp 781–791.
69. Sheckhar R, Evans JW (1990) Fluid flow in pachuca (air agitated) tanks: Part II Mathematical modeling of flow in pachuca tanks. Metall Trans B 21:191–203
70. Shinn JH (1984) From coal to single-stage and two-stage products: a reactive model of coal structure. Fuel 63:1187–1196
71. Stach E, Taylor GH, Mackowski MT, Chandra D, Teichmuller M, Teichmuller R (1975) Stach's textbook on coal petrology. Gebrüder Bornträger, Berlin
72. Stevens SE, Burgess WD (1989) Microbial desulfurization of coal. US Patent # 48511350
73. Stoner DL, Wey JE, Barrett KB, Jolley JG, Wright RB, Dugan PR (1990) Modification of water-soluble coal-derived products by dibenzothiophene-degrading microorganisms. Appl Environ Microbiol 56(9):2567–2675
74. Team of Experts (1994) A strategic document on the biological processing of fossil fuels. Fuel Process Technol 40:379–390
75. Temple KL, Colmer AR (1951) The autotrophic oxidation of iron by a new bacterium: *Thiobacillus ferrooxidans*. J Bacteriol 62:605–611
76. Uhl W, Höne H-J, Beyer M, Klein J (1989) Continuous microbial desulfurization of coal—application of a multistage slurry reactor and analysis of the interactions of microbial and chemical kinetics. Biotechnol Bioeng 34:1341–1356
77. Van Afferden M, Giralt MJ (1994) Desulfurization of dibenzothiophene by bacteria. World J Microbiol Biotechnol 10:510–516
78. Van Afferden M, Schacht S, Beyer M, Klein J (1988) Microbial desulfurization of dibenzothiophene. Am Chem Soc 33(4):561–572

79. Van Afferden M, Schacht S, Klein J, Trüper HG (1990) Degradation of dibenzothiophene by *Brevibacterium* sp. DO. Arch Microbiol 153(4):324–328
80. Whelan PF (1954) Finely disseminated sulfur compounds in British Coals. J Inst Fuel 27(164):455–457, 464
81. Wolfrum EA (1984) Correlations between petrographical properties, chemical structure, and technological behavior of Rhenish brown coal. In: Schobert H (ed) The chemistry of low-rank coals. American Chemical Society, Washington, DC
82. Young KD, Finnerty WR (1991) Molecular genetics of coal-based microbial systems. Workshop # 4 Electric Power Research Institute—US Department of Energy—Second international symposium on the biological processing of coal. San Diego, Ca, USA, pp 4-1–4-54

Adv Biochem Eng Biotechnol (2014) 142: 169–195
DOI: 10.1007/10_2013_197

Published Online: 13 July 2013

Description of Basic Mining Legal Principles

Reinhard Schmidt

Abstract The Federal Mining Act manages access, via the system of mining concessions, to areas free for mining natural resources that do not belong to the surface property and deposits' owner. These cover especially important natural resources for the economy, including coal, ore, salt, crude oil and natural gas, and also terrestrial heat. For mining operations there exist, however, the same decrees for natural resources in the property of the surface owners, which are predominantly higher-value industrial minerals such as roofing slate, basalt, quartz sand, and clays for the fireproofing industry. In the case of mining laws, administrative procedures such as issuing mining concessions, approving operating plans, and issuing permits or licenses to explore according to water rights or the Federal Immission Control Act, those authorities and departments in whose remit the projects fall are dealt with by the Mining Authority. This means that the Mining Authority is the only state point of contact for the applicant, essentially an "all-in-one" service as it will itself instigate any further participation procedures required. The classic licensing procedure of mining is the operations plan procedure, whereby the operator submits an operating plan to the Mining Authority, which then examines it to ensure it fulfills mandatory legal safety objectives. If necessary these safety objectives can be met during licensing of the operating plans by stipulating additional requirements, Depending on the subject and validity period there are overall operating plans having the widest possible remit with comprehensive participation by the authorities and basic operating plans that form the basis for every mining works. There are also special operating plans, which owing to the dynamics of mining, resolve matters that suddenly become necessary or when the basic operating plans as originally conceived were not relevant. The closing-down operating plan is the designated tool for closing down works and for the rehabilitation of the land; in the case of underground mining and mine boreholes an operating history must also be submitted. For those projects that have a significant effect on the environment, an obligatory overall operations plan with mining law project approval procedure and integrated Environmental Risk

R. Schmidt (✉)
Technical University Bergakademie Freiberg, Institut für Bergbau und Spezialtiefbau, Freiberg, Germany
e-mail: karin.kuettner@mabb.tu-freiberg.de

Assessment (UVP) are necessary. The point at which this is required is stipulated in the UVP-mining decree, for example if the mining area of an open-cast pit is more than 25 ha. Alongside the UVP, the procedure is also equipped with public participation and through its "concentrating effect" replaces further licensing procedures according to other laws. The Mining Authority combines supervision and licensing, which are usually inseparable due to the operations plan procedure, as well as aspects of occupational safety and of the protection of the environment. In view of this administrative concentration these should not be fragmented. The "all-in-one" service meets the requirements of a modern public-oriented administration, has only a few points of contact, and can therefore work efficiently.

Keywords Federal Mining Act • Operation Plan Procedure • Environmental Risk Assessment (UVP) • Mining Concessions • German Reunification Act • Mining Authorities • Water rights • Immision Control Act • Mining Supervision

Contents

1 The Federal Mining Act ... 171
2 The Setting Up of and Duties of Appropriate Authorities and Organizations' Legal Framework ... 176
3 Mining Authorities in Germany ... 176
4 Water Rights ... 179
5 Immissions Control Act ... 179
6 Waste ... 179
7 Abandoned Mines ... 180
8 Participation Procedure of Other Authorities and Enquiries from Third Parties ... 180
9 Vocational Training ... 180
10 Mining Concessions ... 180
11 Cooperation Between Mining-Appropriate Authorities and Other Authorities ... 181
12 Operating Plans and Operations Plan Procedure ... 183
13 Basic Operating Plan ... 184
14 Environmental Risk Assessment of Mining Procedures ... 185
15 Mining Supervision ... 187
16 Rights and Duties of Supervisory Technical Officials ... 189
16.1 Mining Authorities ... 189
17 Mining Employer's Liability Insurance Association ... 190
18 Rights and Duties of Mining Companies ... 190
18.1 Responsibilities ... 190
19 Mining Charts ... 191
20 Rights ... 192
21 Duties of Those People Responsible for Safety in Mining Operations ... 192
22 Evaluation ... 194
23 Conclusion ... 195

1 The Federal Mining Act

The legal basis for exploration, mining, and processing of classic mineral deposits in Germany is the Federal Mining Act (*Bundesberggesetz-BBergG*) passed on August 13th, 1980 and newly amended July 31st, 2009. The Act, with its basic decrees, came into effect January 1st, 1982, and replaced the numerous applicable old state acts that had coexisted up until this point. These originated from the period between the middle of the 19th century (1865 in Prussia) and 1978 in Lower-Saxony. Through the enactment of the Federal Mining Act, the fragmented mining laws in the former Federal Republic at that time were subjected to a federal uniform decree.

Significant proven decrees were taken over from the most important previous act, the Prussian General Mining Act 1865 (*Allgemeines Berggesetz- ABG*), including:

- The designated scope for exploration, mining, and processing widened, however, to include reclamation.
- The catalogue of the economically most important deposits.
- The operations plan procedure as a dynamic and pre-emptive licensing procedure in mining, concomitantly as an instrument of works supervision.
- The objectives (safety and health of the employees, prevention of risks to the community, therefore protection of important transport routes from extensive caving, protection of valuable mineral deposits from ruthless exploitation or improper mining) which were to be achieved with the aid of the operations plan procedures, extended, however, to include current concerns such as protection of the environment and waste.
- The mining concessions, a system of mining permits for exploration and exploitation in areas free for mining for mineral deposits, that are not owned by the landowner.
- The system of responsible individuals (formerly supervisors), personal requirements, and delegatable duties.
- Mining damages rights to compensate for unavoidable damage caused by mining activities.

New additions to the act included the following.

- Reclamation, which is an explicit article of the mining law operations plan procedures and not of the universally valid environmental protection laws.
- The extension of the remit to include terrestrial heat and underground containerless storage and disposal as well as (for the time being) underground cavity structures; the latter was resolved in §130, however, it was later revoked.
- The reorganization of mining concessions; the former system of prospecting, requesting a license to explore or to produce, was replaced by a license to explore and license to produce and mine property to exploit in areas free for mining mineral deposits; if the reasons for refusal have not been satisfied or have been dealt with, there is a legal entitlement to be issued permission.

- Data gathering by the states, of field and production output for exploration (according to size) and exploitation (according to quantity) in areas free for mining mineral deposits.
- Authorization to determine zone restrictions.
- The so-called securing of the raw materials clause in §48 paragraph 1 sentence 2: if there are competing public interests of equal importance, mining will be given priority.
- Equal treatment of areas free for mining and mineral deposits of landowners in the property of the surface owner in many areas.

The Federal Mining Act is revised regularly via a range of amendments and articles and brought into line with current legal positions. Only a few important decree regulations have been changed.

- The most important change was the introduction of an obligatory overall operation plan, requiring a project approval procedure with public participation and an integrated Environmental Risk Assessment (*Umweltverträglichkeitsprüfung-UVP*) via the Act of Dcember 2nd, 1990. The inclusion of UVP procedures helped realize European law into national law.

The aim of many European decrees is predominantly the protection of the environment, occupational safety and health protection, and the removal of trade barriers within the European Union. In the field of protection of the environment the above-described implementation of the obligatory overall operation plans with project approval procedure and UVP for those projects was realized, which placed a particular burden on the environment.

The second area, in occupational safety and health protection, the General Federal Mining Decree (*Allgemeine Bundesbergverordnung-ABBergV*) of October 23rd, 1995 came into force that year, and included, among other provisions, the European legal demand for a safety and health protection document with a hazard assessment for every company.

- Other changes of the *BBergG* dealt with deletion of the ambit for construction in rock cavities or tunnel building in the older §130.
- Even outside the limits of mining act project approval procedures, public participation in an "unofficial" operations plan procedure can be authorized by the Mining Authority, if it is foreseeable that more than 300 people will be affected by the project (§48 Paragraph 2).

The Federal Mining Act was preceded by a longwinded co-ordination process among all parties involved (mining trade association, industrial unions of mining and energy), federal departments (economy, work, domestic affairs, and justice), and also the states via the Federation-Federal States Committee for mining, and finally the Federal Council. Among other things, the catalogue of natural resources covered by this law was extremely controversial. The father of the *BBergG*, Dr. Hans Zydek, head of the department in the leading Federal Ministry of the Economy, had initially listed in his draft bill of 1975 all natural resources excluding water. Until then only those natural resources of special significance to the economy were listed in the ambit of the Prussian ABG, and those that did not belong to natural resources in areas free for mining, were listed individually, including coals, ores, and salt, and later crude oil and natural gas. In a Reich's decree of December 31st, 1942, the so-called "New Year's Eve decree," strategically important natural mineral resources such as fireproof clays, fireproof quartz, kaolin, spars, bauxite, diatomite, and other "in the property of the surface owner natural resources" were also placed under mining supervision. The suitability of clays and quartz for fireproof products was set at 1580 °C, so that coarse clay for brick production was no longer included. Essentially, after participation of the industrial associations, the catalogue listed in the Federal Mining Act so-called areas free for mining and in the property of the surface owners' natural resources limited itself to this status quo, with a small exception, that fluorspar and barite were included in the group of areas free for mining natural resources and the catalogue of the property of the surface owners' natural resources extended to roofing slate, basalt lava, and trass (volcanic breccia).

Consequently, the majority of stone, earthen, and therefore building and mass commodities were not mentioned in the individually listed areas free for mining and in the property of the surface owners' natural resources, but as a third category of the so-called "surface property and deposits" natural resources' owner, and as before was not in the ambit of the Federal Mining Act, except underground exploitation.

This marked the failure of the intended comprehensive legal simplification, because of the quantity of shingle and building sands, approximately 250 million t annual production as well as 200 million t broken hard rock produced in Germany, 10 million t quartz sand, and 1.5 million t gypsum, which heavily outweighed production of those natural resources listed in the Mining Act (compare Fig. 1).

The legal position and also the jurisdiction for approval and supervision of surface property and deposits' owners of raw materials vary immensely. The state Industrial Inspectorates as well as the Employers' Liability Insurance Associations are responsible for the supervision of occupational safety, whereas the supervision of environmental protection comes under the aegis of the environment departments, district offices, or regional councils. In addition to this split in the supervisory jurisdiction, the approval system is also confusing to outsiders.

- Surface mining, therefore excavation or a concentration of surface area of more than 300 m^2 and 2 m in height or depth, are considered building constructions according to the Federal Building Code and the state building decrees subject to this code. The authorities responsible for issuing permits are the towns and municipalities.
- If hard rock is mined with explosives or flame jet in quarries, then these works are subject to licensing under the Federal Immission Control Act (*BImSchG*) and 4. *BImSchV*. The relevant authorities are the district offices or municipalities not associated with a county.

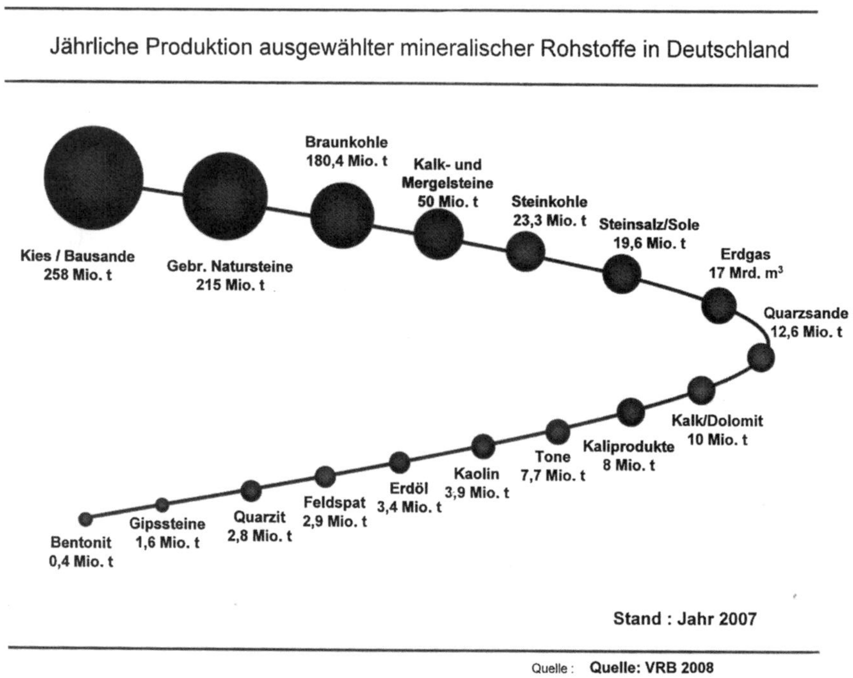

Fig. 1 Annual Production of Raw Materials in Germany

- If gravel or gravel sand is extracted by dredging below groundwater level, for example, with the aid of suction excavators, a permit regarding water rights is required. The district office acts as the lower water board, and for larger projects covered by Environmental Risk Assessment (UVP) either the regional councils or those commissioned by the federal state governments as higher federal state authorities are responsible.
- In North-Rhine Westphalia there is in addition to these laws an Excavation Act, and the District Offices are charged with its enforcement.
- The UVP obligation is in accordance with the UVP laws or decrees of the federal states, which have realized EU law.

At first in the GDR, as in the former FRG, the old federal state laws in all their variety remained in force. However, the GDR Constitution in 1949 and the 1969 GDR Mining Act had determined that all mineral raw materials whose exploitation was of economic importance, irrespective of the property of surface owners, were now public property; the natural resources, however, were eventually only listed because of a decree from August 15th, 1990.

The German Reunification Act of August 31st, 1990 extended the ambit of the Federal Mining Act to include the new federal states, however, providing that the raw materials within the context of the GDR §3 Mining Act and the relevant decrees passed, are natural resources in areas free for mining within the context of the Federal Mining Act. Several others, therefore those unlisted natural mineral deposits are, according to this provision, the property of the surface owner's natural resources in the context of the Federal Mining Act, and were only small mineral deposits limited in size and worth and the smallest of companies. The decree governing conferment of mine property of August 15th, 1990, two months prior to reunification, adequately clarified the catalogue of natural resources. This covered not only coals, oil, gas, salt, spars, and industrial minerals but also nearly all raw building materials in the stone and earthen sectors. Consequently, there were only two categories in the former GDR, as opposed to three in West Germany:

- Predominantly areas free for mining.
- Few were the property of the surface owner's natural resources.

This meant that, contrary to the situation in the former FRG, nearly all natural resources including building and mass commodities in the stone and earthen sectors were considered areas free for mining and the companies were subject to mining supervision. In the background of this provision was the assessment that the building sector after reunification would be an important boost to the economy and the production of the necessary raw materials should not be hindered by unresolved questions of property ownership.

Consequently there was a transfer boom of mining authorizations (mining property permits), which were sold by the *Treuhandanstalt* (Privatization Agency) within the context of the privatization concept of the last GDR government to private investors as well as a large number of applications for new licenses and permits. After quick processing by the mining authorities in the new federal states, production of stone and earthenware materials between 1991 and 1995 in some cases tripled. Production figures subsequently fell once more due to the building recession and competition between newly formed powerful companies increased.

One consequence of the different mining laws in West and East Germany were the countless legal cases heard at the highest level as well as initiatives from parties and federal state governments because of a considerable clash of interests, in particular of surface property and deposits' owners in the East, especially the Farmers Association, which did not own any mass commodities, and the municipalities (Cities and Municipalities Conference) who should have been responsible for enforcing this act according to building laws, which henceforth, however, were enforced by the federal state mining authorities.

This process was concluded by an Act to simplify the legal status of mineral deposits (*Gesetz zur Vereinheitlichung der Rechts-verhältnisse bei Bodenschätzen*, April 15th, 1996) which could not rescind the German Reunification Act, but prevent its future application on this issue. For constitutional reasons, all permits issued until this point were granted a right of continuance, so that the vast majority of the companies' (ca. 80 %) overall production (ca. 95 %) remained under mining supervision and only new projects and the few formerly in the property of the surface owners' natural resources that according to the new Act were too small.

In view of the fact that in the former GDR nearly all the stone and earthen mining is under the mining authority, only the data gathered here are accurate.

2 The Setting Up of and Duties of Appropriate Authorities and Organizations' Legal Framework

As concerns mining supervision, the Federal Mining Act states in §69 paragraph 2, that mining is subject to supervision by the appropriate authority (mining authority).

In §142 it is stated that the federal state governments or those authorities designated by them to implement this Act, themselves appoint the appropriate authorities. The implementation of the Federal Mining Act is therefore realized differently by the federal states.

3 Mining Authorities in Germany

In the postwar period in the three occupied zones in western Germany, mining along with agriculture was the most important sector for rebuilding the country after being defeated in the war. The Western allies favored federalism when politically rebuilding Germany, as it had a long tradition and had only been completely eliminated during the period of the Third Reich from 1933 to 1945. In 1947 the federal states, then in 1949, the Federal Republic of Germany and, in the Soviet occupied zone, the GDR were founded. The mining authorities had up until this time (except under the Third Reich) always been state authorities; their tradition dated back to the medieval concessions of the mineral royalties. The oldest available document concerning the office of a so-called Master of Mines comes from Freiberg, dated 1241.

With the support of the occupying powers, the mining authorities were re-established according to a federal pattern. In the main it involves a three-tiered administrative structure with the Supreme Mining Authority (Department of Trade and Industry), and the higher (Chief Mines Inspectorate) and lower administrative levels (Mines Inspectorate). City states or states with little mining acceded via treaties to the Chief Mines Inspectorate of neighboring federal states, for instance, the Chief Mines Inspectorate in Clausthal-Zellerfeld acted on behalf of the federal states of Lower Saxony, Schleswig–Holstein, Hamburg, Bremen, and initially even West Berlin. In the Soviet-occupied zone and eventually in the GDR, Mining Authorities were disbanded and transformed into Technical Mining Inspectorates, which basically had to yield to the goals of the centrally planned economy. After numerous serious mining accidents in the GDR, a designated authority for mining supervision was set up once more in 1959. It included a Supreme Mining Authority in the Council of Ministers of the GDR in Leipzig, which at the end oversaw six Mining Authorities.

After reunification, despite early efforts to form cross-national Mining Authorities—for example, in Lower Saxony—due to the growing self-confidence of the new federal states this was only possible between Brandenburg and Berlin. Initially the federal states of Saxony, Bran-denburg, and Thuringia founded their

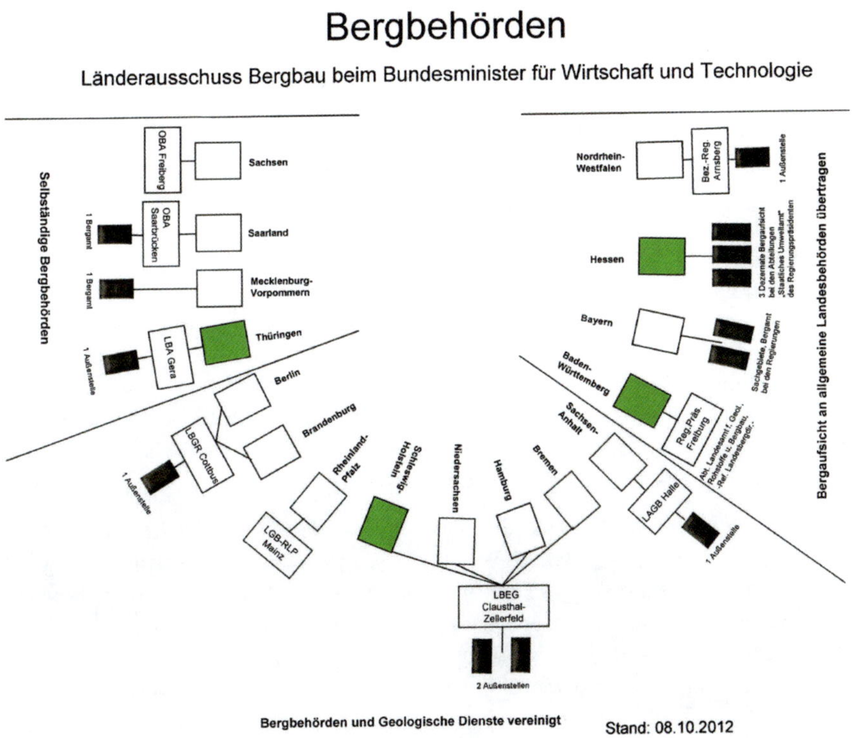

Fig. 2 Mining Inspectorates, Federal Government—State Committee for Mining at the Federal Ministry for the Economy and Technology

own Chief Mines Inspectorates and Mines Inspectorates, which were in part later restructured in several stages via administrative reforms in order to reduce personnel. The current situation of the Mining Authorities of the federal states, whose binding element is the Federal Government–Federal State–Conference for Mining at the Federal Ministry for the Economy and Technology, is illustrated in Fig. 2.

The Supreme Mining Authority is always the Department of Trade and Industry or the Economics Senate with the exception of Hesse and Thuringia, and recently also Baden-Württemberg and Schleswig–Holstein, where the Environment Ministry is the Supreme Mining Authority.

The main task of the Mining Authorities is the formulation of natural resources and energy political goals of the federal state governments. The Supreme Mining Authorities (which, along with the regional councils actually belong to the middle tier, but are responsible for the whole state, and are therefore described as "higher" authorities) formulate mining decrees and technical guidelines of general significance. In addition to this they issue mining permits for areas free for mining natural resources and carry out extensive official administrative procedures. The lower Mining Authorities are responsible for directly overseeing the companies and the more basic irregular licensing procedure, especially the operations plan procedure.

In the last ten years, the following changes have occurred at the middle level.

- Two tiers (excluding Saarland) meaning that beneath the Ministry level there is only one statewide competent authority (if necessary with branches), which combines the tasks of the higher and the lower administrative levels, for example, in Thuringia and Saxony, and temporarily as an intermediate stage in Lower Saxony and Brandenburg,
- Merging the Mining Authorities with the geological services of the federal states, for example, in Brandenburg–Berlin, Saxony–Anhalt, Lower Saxony, Rheinland–Pfalz as an intermediate stage, and also temporarily in Baden–Württemberg,
- Closing the Mining Authorities as special authorities and incorporating them into regional councils, for example, in Bavaria, Hesse, and Baden–Württemberg, and in North-Rhine Westphalia.

The background to this total disbandment was due in part to the insignificance of mining in Bavaria or Baden–Württemberg. In Hesse and North-Rhine Westphalia it was a result of state politics. Merging mining and geology, however, combines different modes of operations: implementation by the Mining Authority and predominantly scientific work by the geology department. Both are responsible for securing natural resources, which in view of the current world market, has been given a totally new perspective. Therefore this merger is a sensible step, reducing bureaucracy while maintaining the independence of these authorities.

The cooperation of several federal states on the basis of treaties, as adopted in Lower Saxony, Brandenburg, and in former times in the Saarland together with Rheinland–Pfalz, is considered beneficial by all experts.

The incorporation of higher authority responsibilities into a general administration cannot adequately satisfy current demands of securing natural resources. Furthermore, far-reaching decisions in a general administration are normally made according to the current political climate. Natural resources-relevant decisions require a long-term perspective and independence when considering the different interests, and must not be screened beforehand within the same authority, for example, such as environmental and regional development.

The oldest mining administration of Germany in Freiberg Saxony kept the name *Oberbergamt* (Chief Mines Inspectorate) for historical reasons. It has been retained in the new two-tiered mining administration because the description dates back to 1542 and was a nucleus for many other institutions. The founding of the University of Applied Sciences Mining Academy Freiberg in 1765, the oldest mining university in the world, can be traced directly back to the director of the Chief Mines Inspectorate. Its predecessor, the *Stipendienkasse* of 1702, was the oldest state teaching establishment for mining worldwide. In the nineteenth century, the state geological office was created by outsourcing the department *Geognostische Landesuntersuchung* (land exploration). Even the term *nachhaltigkeit* (sustainability) was coined by the Freiberger *Oberberghauptmann* v. Carlowitz 1713 (Chief Mines Captain) and originates from the Chief Mines Inspectorate.

4 Water Rights

In addition to these basic responsibilities designated in the Federal Mining Act, the Mining Authority is responsible for the implementation of the Water Resources Act (WHG) in mining facilities whereby the Mining Authority takes the place of the Water Authority. Licenses to explore are nevertheless only to be issued with the consent of the Water Authority. Typical examples are licenses to explore the use of groundwater before lowering the groundwater level for open-cast pits or discharging water from purified processing water into public waters and the planning approval of open-cast pits for dredging below groundwater level while exposing strata where the groundwater flows.

5 Immissions Control Act

Additionally, the Mining Authority is responsible for the implementation of the Federal Immissions Control Act (*BImSchG*) in works subject to mining supervision. This covers predominantly irregular permits for setting up and operating processing plants, for example, crushing and sifting plants, and also regular permits, for example, for rotary kiln units to clean contaminated ground within the remit of rehabilitating lignite processing facilities.

6 Waste

There is an exception clause in the Act for Promoting Closed Substance Cycle Waste Management and Ensuring Environmentally Compatible Waste Disposal, in favor of the Federal Mining Act when unavoidable mining waste is created during exploring, mining, and processing mineral deposits. According to this Act, waste is to be disposed of without any danger to the public within the framework of mining law operating plans. These can be used for a variety of mining-technical purposes because of their geotechnical properties.

Concomitantly it is possible to use nonmining mineral waste for these applications. Recycling mineral waste ensures, according to a ruling of the European Court in 2003 and of the Clay Quarry ruling of the Federal Constitutional Court in 2005, a totally adequate waste recycling within the meaning of the Waste Act. Many mines are used throughout the FRG to dispose of waste, also regulated by the WHG.

7 Abandoned Mines

When the reclamation works in the closure plans have been completed and the works are longer expected to present a danger to the life and health of a third party or the general public, mining supervision ends according to §69 Paragraph 2 and in the event of any subsistence is not renewed.

In such cases concerning historically abandoned mines without any legal successor, the Chief Mines Inspectorates becomes a "Special Regulatory Authority" by the states and is therefore responsible for public safety and order in abandoned mines and other underground cavities. Because public safety and order are state responsibilities, the Mining Authorities award contracts to secure and reclaim affected areas.

8 Participation Procedure of Other Authorities and Enquiries from Third Parties

The Mining Authority represents the concerns of securing natural resources and public safety in official statements concerning planning schemes and licensing procedures of other authorities acting as a public welfare agency.

9 Vocational Training

The Mining Authorities are responsible for both the work experience qualifications before and during mining studies as "hardworking miners" as well as probationary service in the higher civil service in mining and mine-surveying as mining trainees/mine surveyor trainees after completing their university education. In addition to North-Rhine Westphalia and Lower Saxony, Saxony because of its close links to the Technische Universität Bergakademie Freiberg, also belongs to the leading states (in terms of education and training) in Germany.

10 Mining Concessions

For exploration in areas free for mining natural resources, listed in §3 Para. 3 *BBergG*, those raw materials of special importance to the economy such as coals, ores, salts, crude oil, and natural gas, according to §6 at the place already sited, a license to explore and a license to produce for the production or mining property (lordship) are necessary. The mining concessions are allocated irrespective of the surface and deposits' owner. The mining company is obliged to seek consent of the owner of the surface and mineral deposits when wishing to mine on land they do not own.

A license to explore (for exploration) and a license to produce (for production) must be issued if the requirements for a refusal listed in §§11 and 12 *BBergG* are not met. Therefore the applicant has a legal entitlement, but the issuing authority has only limited discretionary powers. A license to produce can be converted into mining property (lordship), resulting in essentially the same rights, but easier to borrow against, over a longer term, from the banks.

In order to ascertain rights of ownership of apparently "unowned" areas free for mining mineral deposits, a license to produce must be applied for from the Mines Inspectorate. This application must list the "prospective" natural resources as well as the location and size of the area for which a license to produce is being applied. Additionally, the proposed technical procedure of the development and production are to be outlined in a work plan. The work plan is of particular importance in ascertaining primacy among competing applications, whereas the date of application is immaterial to precedence. The issued license to produce is a state concession for the appropriation and extraction of the natural resources and protects the holder from competition. It grants in disputes with the surface property and deposits' owner the right to demand *Grundabtretung* (acquisition procedure). This constitutes compulsory purchase in the event that an amicable agreement cannot be reached, if the raw materials extraction is in the public interest, for example, in supplying the economy with natural resources or preventing redundancies.

The goal of a license to produce is the extraction of the raw materials and not hoarding permits in order to eliminate the competition. For this reason the license to produce is only valid for three years, during which an operating plan to commence procedures must be submitted. If this is not adhered to, then the license to produce expires. If the stated extraction goals, submitted in the operations plan are achieved (i.e., buildings constructed, plant operated), the license to produce is normally limited to 50 years, with the possibility of an extension.

11 Cooperation Between Mining-Appropriate Authorities and Other Authorities

According to §15 *BBergG*, the Mining Authority before considering an application (in this case for mining concessions) must give the authorities an opportunity to express their views, whose responsibility it is to ensure that public interests are fully realized. For mining concessions procedures these were the regional councils and the administrative districts. The legislator does not explicitly stipulate the participation of the municipalities, which in Saxony in order to keep the peace nevertheless have, per departmental decree, voluntarily been included in the procedures since 1994.

The duty to ensure participation of other appropriate authorities for their respective sectors, but also for other administrative acts of the Mining Authority, is set in law. In particular the following has been established in the operations plan procedure, according to §54 Paragraph 2: if through the measures intended in an operating plan the remit of other authorities or the municipalities as planning agents is infringed, these must then be included in the decision-making process by the appropriate authority before the operating plans are approved. The participation of third parties has not been allowed for in the Federal Mining Act, although they can be included according to the Administrative Procedures Act §13 Paragraph 2 sentence 1, if their legal interests can be affected by the outcome of the procedures. In Saxony organizational participation is mandatory as per §57 of the Saxonian Nature Conservation Act and §60 of the Federal Nature Conservation Act.

In the event that more than 300 people are affected, §48 Paragraph 2 *BBergG* also allows for public participation in irregular procedures (operating plans). This right (of the decree), which had been incorporated into the Federal Mining Act since the Moers–Capellen ruling of the Federal Constitutional Court of 1989, was first exercised in the federal state of Saxony, in the case of an operating plan to raise the groundwater level within the course of lignite rehabilitation by LMBV.

In addition to "participation" in the procedures of the Mining Authorities, an "understanding" according to the Water Resources Act is necessary, which essentially means official approval must be obtained from the water authority. If this cannot be achieved, an agreement has to be reached on a higher level (Mayor's Office/Regional Council or Department of Trade and Industry/Department of the Environment).

A peculiarity of cooperating with other authorities is found in the "securing of natural resources" clause in §48 paragraph 1 sentence 2. When implementing statutory provisions care must be taken that exploration and extraction are impeded as little as possible. In practice this means that mining will be given priority in a situation of conflicting public interests; when considering pros and cons mining must be prioritized. This also means for the other authorities, that they have to assume mining interests will override others, if no predominantly public interests militate against or even forbid mining (§48 Paragraph 1 sentence 1), for example, on land that as per decree has been officially dedicated for public purposes. Among other things, these decrees take into consideration that a mining project is bound geographically to the mineral deposit, as a result of which there is significantly less planning freedom than in other projects in trade and industry.

A hallmark of the "lignite" states of North-Rhine Westphalia, Brandenburg, Saxony–Anhalt, and Saxony is the lignite planning procedure, which is carried out at the same time as the mining law operations plan procedure. In contrast to the operations plan procedure, the lignite plan is not drawn up by the mining company, but by the Lignite Planning Office of a regional planning association. They focus particularly on land use planning concerns. These factors come together in a lignite plan, which combines the varied public interests of the planning region. The plan is passed in an "association meeting," presided over by the district administrators of the planning region, and then approved by the supreme appropriate authority for land use and planning, the Ministry of the Interior.

As all lignite mines in Saxony had already been approved before the *BBergG* came into effect on October 3rd, 1990, no mining project approval procedure with UVP could be carried out due to the provisional regulations, which federal and state higher courts later confirmed. From a mining law perspective these companies are operating on the basis of general operating plans, and the Mining Authority required the submission of facultative overall operation plans. For this reason the lignite plan is of particular importance as a vocalization of public interests in the region.

The Mining Authority, under its own basic responsibility, safeguards all aspects of occupational safety and health protection. Participation of Industrial Inspectorates or the Employers Liability Insurance Associations is therefore not intended in the licensing procedure. However, there is cooperation, in particular with the Employers' Liability Insurance Associations and also with the business associations and unions, as social partners, when drafting technical rules or mining decrees.

12 Operating Plans and Operations Plan Procedure

The mining law operations plan procedure is a centuries-old instrument. During the period of the directive (directorial principle) it also served to control company constraints. It gradually developed into an inspective (inspectorial principle), reducing the remit of mining supervision to policing duties, such as safety, health protection, surface land conservation, and ensuring public welfare, and was retained for well-considered reasons in the mid-nineteenth century. The official explanation for its retention in the Prussian ABG of 1865 clearly lists benefits that have not changed to this very day:

- An orderly business can only be directed on the basis of a plan.
- The prior submittal of this (operating) plan to the Mining Authority as a license application has not significantly increased bureaucracy.
- It combines, however, approval action with supervision by giving the Authority via the operating plan detailed insight into the company, making frequent inspections no longer necessary.
- A prior submittal of the operating plans lies in the interests of both the company and Mining Authority.

The operating plan, a typical mining permit instrument, was also integrated into the Federal Mining Act of 1980 which because of its dynamic operating mode could respond immediately to unforeseen occurrences at a mining company.

Approval of the operating plans is linked to a catalogue of "protection goals," commensurate with the specific responsibilities of the Mining Authority: occupational safety and health protection, providing for environmental protection, securing mineral deposits within the remit of relevant procedures, and ensuring public welfare. On the other hand there is a legal entitlement to be given approval if these criteria are met, therefore the authority has only limited discretionary powers.

Subject to the matters of control the legislator has created several kinds of operating plans in §§52 ff.

13 Basic Operating Plan

The legal basis for running a mining company is the basic operating plan (§52 paragraph 1). A basic operating plan, from which the actual permitted activities of the company emanate, is necessary in each case in order to exercise the right conferred by a mining concession (license to explore/license to produce/mining property, "lordship") in areas free for mining mineral deposits, or on the property of the surface owners' natural resources (obtained via purchase or contract to use the property). It is normally valid for two years, therefore the procedure corresponds to the dynamic operating mode of the mines whereby the number of miners varies constantly according to actual mining progress.

The legislator has designated the following types of operating plans, from the setting up to operating to closing a mining operation, in the following order.

1. **Overall Operations Plan**

The overall operations plan considers the large-scale parameters of the projects, when necessary in competition with other public and private interests and serves in particular to bring about a wide participation as early as possible, of authorities and departments and sometimes also associations. Preferably it should stay valid for a long time, thereby allowing for the setting up, operating, and closing down of new projects. This ensures the effects on third parties and also the environment are transparent.

1.1 The facultative overall operation plan, already stipulated by the legislator in the first draft of the Federal Mining Act, is to be presented to the Authority upon request. In practice it serves, among others, uniquely to enable as wide a participation as possible, thereby simplifying, in this respect, the subsequent procedure of the basic operating plans, from which as described the actual permitted activities emanate.

1.2 In the widest reaching amendment to the *BbergG*, the legislator introduced the obligatory overall operation plan in 1990. This must be submitted by projects that present a particular environmental challenge. The "relevance threshold" of the "challenge" is formulated in the UVP-Mining Decree of 2005, for example, by open-cast pits which are currently a mining area of over 25 has; Germany followed EU guidelines and the modus operandi of the other member states. From 1998 to 2005 the threshold of over 10-hectares mining area for open-cast pits was far more restrictive, and at its most extreme before 1998, a total works area of over 10 ha.

The obligatory overall operation plan is to be submitted before beginning the projects and must also cover setting up. Approval may be given in stages for very practical reasons; because of the highly complex procedures and the time required there is the option of starting prematurely. Such a case would be when a positive response, due to the submitted documentation, can be expected, the projects are in the public interest, and the applicant has agreed that in the event of a refusal, to restore everything to its former condition.

2. Next in the line is the basic operating plan, which covers the actual execution of the work to be carried out (already described above).
3. For individual schemes that suddenly become necessary and could not be addressed in the basic operating plan, there exists the instrument of the special operating plan, which to a significant extent takes into account the dynamic operating mode of mining. The special operating plan is not limited in time, but in scope, that is, to a single scheme. Possible schemes falling within this remit are, for example, erecting a fire dam around a fire area, the re-driving of a dropped part of a drift, or inspection of a conveyor belt or a ramp.
4. For projects that have several operations, such as the siting of a central stockpile for a common coal mix and storage yard or a common mine drainage station, there is the option of joint operating plans.
5. Closing down the operations, which entails ceasing production and reclaiming the surface area used, must be outlined in the closing-down-plan. The plan remains valid until official supervision of the mine ends, therefore when it can be expected that the site/operation no longer presents a hazard. All reclamation procedures must be completed. The end result must be usable "blank" land.

Rehabilitation does not automatically mean restoration of the former condition, but the orderly landscaping of the land surface affected by the mining, keeping in mind public interests (§4 paragraph 4 a. a. O.). It is quite possible that public interest, for example, as formulated by the District Offices as the lower environmental protection authority, would prefer to see a lake in a disused open-cast pit hole instead of the former agricultural land or woods, or using a former clay pit as a landfill site.

The closing-down plan of underground mines or boreholes must be accompanied by an operations history, which gives future generations information about the remaining mineral deposits, quality characteristics, and intended purpose, and particularities during processing, making it easier to decide whether to reopen the mines if the economic climate has changed.

14 Environmental Risk Assessment of Mining Procedures

Along with the obligatory overall operation plan described in above 1.2, an Environmental Risk Assessment (*Umweltverträglichkeitsprüfung-UVP*) according to §57a paragraph 2 and §57c is necessary, which must consider the effects of the project on the environment. The environmental risk assessment consists of a macroscopic (state planning) part, which usually is granted from the appropriate authorities for regional development and state planning in the form of a precursory nonindependent regional planning procedure. In Saxony these are the regional councils. In the microscopic part the results on the basis of official statements of the authorities involved are to be considered by the Mining Authority accordingly and integrated into the decision. The final decision is made by the planning approval authority, the Mining Authority. In several federal states this responsibility was occasionally realized by the highest authority, the Department of Trade and Industry.

The procedure itself is a mining laws project approval procedure. As such it has a concentrating effect, therefore all concerns (excluding nuclear) are considered in one official statement. This also encourages an intermediary examination, ensuring that the interaction of different factors is taken into consideration; consequently, unlike the case of the normal irregular operations plan procedure along with an operating plan permit, for example, there is another *BImSch*-permit for water treatment or a water authority license to explore to tap the groundwater in order to drain it.

The disadvantage of an insufficient concentration effect in a normal operations plan procedure is offset by speed and adaptability. To balance this out the Mining Authorities of the federal states have been given special responsibilities for immission control, water, and waste, which are clearly defined, also helping to encourage an intermediary examination without, however, fulfilling the demanding formal requirements of the project approval procedure.

In addition to this the project approval procedure is linked to public participation. After participation of the authorities, as a result of which the application documents can be changed, the projects are made public in the media and put on display for public inspection, for example, in the town hall. Private individuals who feel this project infringes upon their rights can raise their objections within the predetermined period. In a hearing, to which the person raising the objections, applicant, technical authorities, and when necessary an expert are all invited, these objections can be discussed and if possible also eliminated, with the results then incorporated into the planning approval decision.

On occasion procedures other than the mining law procedure can be UVP-binding. These include official procedures according to the Federal Immission Control Act, as long as these projects are listed in the catalogue of the 4. *BImSchV* in column 1; the projects in column 2 of this decree require only an irregular procedure.

Project approval procedures according to water and waste legislation, which are also UVP-binding, can also be linked to mining projects. Often they are, but usually once mining has ceased. If, for example, after production has ceased in a large lignite open-cast mine, and the remaining hole should be formed into a permanent lake, project approval procedure according to the Water Resources Act (*Wasserhaushaltsgesetz-WHG*) is necessary. The actual planning approvals authority according to water rights varies in the federal states. In Brandenburg, for example, the Mining Authority is also the planning approvals authority according to WHG in the above case. If a landfill site is to be opened in an open-cast pit after production has ceased, this would require under certain circumstances UVP as part of the waste legislation project approval procedures.

Scoping meetings are an additional advance Authority service to coordinate the approvability of a project with the authorities and the applicant before procedures officially begin. At this point it is also clarified which documentation is to be submitted with the plan.

15 Mining Supervision

Mining supervision is carried out, as already described, by those offices designated by the federal state governments. Since the mid-19th century when the inspection principle was taken over from France, this has included reducing the policing function duties, especially employee health and safety protection and protection of the environment, as well as securing natural deposits.

Responsibility for operating safety lies strictly with the mining company, and is subject to the control of the supervisory authorities. In practice the above-mentioned safety objectives are achieved within the scope of state implementation. The mining authority/mining supervision covers the following aspects.

1. **Supervisory Function**

The supervisory function is performed by the higher and executive mining technical office of the Mining Authority. These are university-qualified mining engineers and in addition to this in several federal states with extensive mining, also mining inspectors, usually experienced staff or technicians who carry out works inspections. The mining inspectors are instructed by a social-political advisor, who is a trained engineer and realizes the specific concerns of work safety. He is also the contact person for the works council member of the mining works.

2. **Permits**

As already discussed in operating plans, granting permits is a further important aspect of state duties. The peculiarity in mining is that, as described previously, the operations plan procedure as a classic licensing procedure of mining is also an effective instrument of mining supervision, because through the submission of the operating plans detailed works information and descriptions are given directly to the Mining Authorities. In mining legal terms, supervision and granting concessions are therefore inseparable, which is often stipulated from outside in particular by environmental authorities and classic administration departments from the Mining Authority.

One result of the special remit is that in mining there are usually also licensing procedures according to water, immissions, and waste rights all in the hands of the Mining Authority, which then involves the appropriate subject office. For water rights procedures it also ensures an "understanding" is reached. The mining company is offered an all-in-one service, conforming to the current administration's postulate of "one address." The European Union called this German system "Best Practice."

Outside of mining projects and mining supervision, which cover licensing procedures according to different legislation (building laws, *BImSchG*, WHG, surface mining act, waste industry act, and the circulation act), such a unity of supervision and approval, does not exist. Licensing authorities are usually the district offices as the lower, or the regional, councils as the higher immission control, water, or waste authorities, or the Council Building Authorities. The State Industrial Inspectorate is responsible for supervision and, irrespective of this, the Employers' Liability Insurance Associations for occupational safety, and the State Environmental Departments.

A unity of supervision and licensing is therefore only guaranteed by the Mining Authorities. The environment authorities often criticize mining supervision, because this takes place outside their direct authority. The Mining Supervision's assessment criteria of the protection of the environment are, however, the same as the environment authorities': they are laid down in the relevant subject laws as well as in the technical rules and regulations of LAGA (federal states working group waste) and LAWA (federal states working group water) and are applied similarly. One difference is, however, that supervision by the environmental authorities initially ignores occupational safety and health protection.

3. **Regulatory Jurisdiction**

In addition to the listed responsibilities of supervision and licensing, a part of the regulatory jurisdiction lies with the Mining Authorities. Mining has specific dangers due to its dynamic operations, such as rock burst, firstfall and slumping, water penetration, gas, dust, and danger of burning or explosion. Consequently the specific rules and regulations of the mining decrees, and not the accident regulations of the Industrial Employers' Liability Insurance Associations apply. These are in part passed by the Federation (BMWT), and in part by the federal states. The authorization by the Federal Mining Act includes the decree of the higher Mining Authorities. The first priority of the Mining Authorities is to ensure an orderly and safe operation of the plants and approve the operations plan via commitments to generally recognized technical rules. In the states there are special mining decrees for individual mining branches, for example, hard coal mining, lignite mining, ore/spar, and abandoned mines as well as crude oil and natural gas exploitation. Decrees of general importance, such as for hoisting plant and inclined conveyors, and the electrical mining decree, are harmonized as far as possible in the federal states via the federal states mining committee. They are therefore passed in conformity, which is of particular importance for cross-border mining. Industrial medicine is also covered by decree.

16 Rights and Duties of Supervisory Technical Officials

16.1 Mining Authorities

Responsible companies and individuals have a duty of disclosure to the technical supervision officials of the Mining Authorities and the obligation to submit documentation when requested.

The individuals charged with supervising mining by the Mining Authorities are authorized (§70) to enter works premises, offices, and facilities of the respondents, as well as watercraft (e.g., a floating dredger), carry out tests, inspections, and take samples with them, and are allowed to inspect the business and production documentation.

In order to prevent imminent danger to public safety and order, the above-mentioned land and premises may also be entered outside normal company working hours and also when they serve as accommodation as well. This limits the basic right of inviolability of the home (Art. 13 *Grundgesetz–GG*) accordingly. The latter applies especially when investigating accidents, in which case the technical members of the Mining Authorities are auxiliary officials from the public prosecutor's office. In addition to this, the relevant Authority can sanction specific measures in individual cases, in order to protect life, health, and property of employees or third parties when necessary. The authority of the decree is also necessary, in the event of an operation being closed down without an authorized closing-down procedure plan and also in an emergency, if no responsible party can be found, to ensure the above-mentioned requirements are met.

In the event of unapproved activities, exploration and extraction in areas free for mining mineral deposits without a concession or running a company without an operating plan or permit, members of the technical service of the Mining Authorities can stop production, and in the field of the continental shelf even decree that all facilities be removed. Explosives and ignition charges can be secured and used in the event of a missing permit or unauthorized use.

The Mining Authority can forbid an operator employing a responsible employee in the field they have been assigned to, if they deliberately or with gross negligence contravene their responsibilities, or other facts come to light and misgivings arise about the necessary reliability, expertise, and physical suitability. If an operator does not respond to this order, all production can be stopped. This also applies if misgivings arise as to the operator's reliability.

The Mining Authority can order action to be taken to save injured people or those in danger in situations where there is a danger to both employees and third parties. Normally the rescue system is the responsibility of the operations manager in co-operation with the Chief Rescue Officer of the mine rescue brigade.

The Mining Authority takes the place of the Industrial Inspectorates officials (who are only responsible for other industries) in the fields of technical and organizational safety of operations, and for ensuring health protection.

17 Mining Employer's Liability Insurance Association

The technical supervisory officials of the Mining Employers' Liability Insurance Association (*BGRCI-Berufsgenossenschaft Rohstoffe, Chemische Industrie*), in contrast to other sectors in mining, do not have a supervisory or directive function. They work in a consultative capacity in the fields of occupational safety and health protection, as well as the prevention of occupational illness, accident recuperation, rehabilitation, compensation, and annuities payment to bereaved families and occupational illness of deceased insurants. In addition to this the BGRCI maintains research institutes and clinics. The Mining Employers' Liability Insurance Association also organizes and maintains central offices for mining rescue in Clausthal-Zellerfeld, Hohenpeißenberg, and Leipzig; the central offices Friedrichsthal (Saar) and Herne (Ruhr) are maintained by the German Hardcoal Authority (*Ruhrkohle AG–RAG*) and subject to the Mining Authority.

In the former GDR there are also many building raw materials extraction works subject to the Mining Authority. The Quarry and Underground Mining Employer's Liability Association is the institution responsible for the social pension fund and accident insurance, which in these particular cases also do not have a supervisory or directive function. They nevertheless work in close cooperation with the Mining Authorities in a consultative capacity.

18 Rights and Duties of Mining Companies

18.1 Responsibilities

Since the Federal Mining Act came into force, the responsibility for orderly management as well as safety and order in the operations is the responsibility of the mining company. This includes, in particular, adherence to generally recognized rules of safety equipment, even if these are not official legislation, as well as other accepted work, scientific expertise, industrial medicine, and work-hygiene rules (§61).

The operator may delegate part of his responsibility, such as managing the operations or processes to other responsible individuals. These individuals must receive written appointment and the Authority notified. Prerequisites required to carry out these duties by the responsible individual are their reliability, expertise, and physical suitability. All these are to be verified, for example, by a copy of criminal records, certificates from a college/university, and a certificate of mining suitability from a doctor qualified to do this.

When the operator delegates a responsible person to manage the works, his own personal "reliability" suffices. The operator is also responsible for circumstances or in situations that could pose a direct threat to the life or health of employees or third parties, and at his own expense must take suitable action to alleviate danger or rescue the injured. This also applies to accident assistance agreements with mine rescue brigades from nearby operations and the appropriate equipment.

19 Mining Charts

The operator must maintain mining charts for every extraction works and underground explorations- operations, kept up to date within the mandatory time period. Mining charts include an overview of the mine, other information such as outlines, maps, and plans. The mining charts must be drawn up and completed by a mine surveyor (*Markscheider*) recognized by the Mining Authorities. The mine surveyors are graduate mining land surveyors with a university degree. When applying their expertise they are completely autonomous and enjoy, like notaries, public faith.

In contrast to many of the mining regulations of the federation and the federal states, which each had individual safety aspects dealt with in detail in the various mining branches, the General Federal Mining Decree (*ABBergV*) should be viewed as a covering law for safety and health protection in mining. It therefore replaces a range of old mining decrees in the old federal states of Western Germany. In the eastern federal states a decree will be issued that supersedes some of those time-limited (in part due to the German Reunification Act till 1995), yet still applicable former GDR mining decrees.

The background to the *ABBergV* decree is the implementation of European law into national law, in particular the (framework) directive 89/391/EEC concerning the implementation of measures to improve safety and health protection from July 12th, 1989, and a further 13 individual directives, for example, for drilling wells, underground works, use of working equipment, protective clothing, safety markings, handling of loads, and visual display units.

The *ABBergV* emphasizes the personal responsibility of the operator for safety and health protection and codifies the responsibilities and rights of the employees. A particular focus of the *ABBergV* is the introduction of the safety and health protection documents, which have hazard assessment at their core. This hazard assessment is to be carried out in three stages: ascertaining the dangers, assessing them, and having a written record of the results of the assessment.

In addition to the technical measures ensuring safety and health protection in the works, a range of organizational measures has also added to the responsibilities of the operator in the *ABBergV*.

- The operators have to take into consideration specific dangers, accident prevention, and contingencies in case of emergency.
- The design of the workplace, bearing in mind the natural conditions, must conform to basic ergonomic and safety standards guidelines. This includes choice, use, and maintenance of working equipment, personal protective equipment, and health and safety protection marking at the workplace in those

cases in which the risk or dangers cannot be avoided just by technical or organizational measures alone.

20 Rights

In order to ensure the highest possible extraction of the deposit, the mining company has the right:

- For the land it requires, whose owners are not prepared to sell or sign a utilization contract, to demand acquisition procedure according to §77 ff. *BBergG*. This is bound to an official administrative procedure via the Mining Authority. In practice, an acquisition procedure is considered when setting up open-cast pits on conferred areas free for mining natural resources or when setting up necessary surface installations, for example, a shaft site for underground mining. The surface property and deposits' owner has to tolerate underground mining beneath his property and has no right of appeal (German Civil Code *Bürgerliches Gesetzbuch–BGB* §§905 and 1004).
- To ensure maximum extraction of the deposit, the mining company has the right, outside their field, to construct auxiliary drifts on fields not their own, for example, for mine drainage or ventilation, use mine openings not their own for the above-mentioned purposes, and demand *Zulegung,* thereby also extracting natural in neighboring fields that would otherwise not have been extracted.
- The mining company can use the provisions concerning the "concessions without latitude" mining concessions and operating plans to their advantage. In the event of absence or elimination of impediment or reason for denying approval they have a legal entitlement to a concession.
- Within the scope of the securing natural resources clause in §48 paragraph 1 sentence 2 *BBergG*, the mining company may assume that in the event of a conflict of interests the Mining Authorities will decide in favor of mining.

21 Duties of Those People Responsible for Safety in Mining Operations

As mentioned above, the mining company operator is responsible for safety and order in the works. The option of delegating part of this responsibility does not absolve him or her from this obligation, to ensure that those duties delegated to the "responsible persons" are unambiguous, cover all salient points, and have a clearly defined remit. In addition to this the responsible persons must be provided with all necessary documentation and operating plans.

If the Federal Mining Act primarily addresses the operator, the General Federal Mining Decree (*ABBergV*) also has decrees governing the actions of the responsible persons and the employees. Form and scope of the supervision by the responsible person have been consolidated in the *ABBergV* insofar as that somebody must be responsible for every occupied workplace. Inspection frequency and regulations governing the individual workplace are all clearly defined.

"Responsible persons" must not only be able to fulfill their obligations, but organizational measures within the company must guarantee that all obligations are met.

If employee representation in the company is not mandatory according to the Works Act, the individual employees have the right, according to *ABBergV* §7, to be heard.

Only those procedures may be delegated to them that are commensurate with their knowledge, experience, and physical abilities. Dangerous procedures may only then be carried out when a "responsible person" has authorized this. There must be a written procedure and those employees concerned must be acquainted with it.

The employees are to be informed immediately about specific dangers during emergencies and also about protective measures. As protective measures in a company often do not suffice, the General Federal Mining Act also lays down that the employees must behave in a safe manner. They have as far as possible to look out for their own safety and for the safety of others who could be affected by their actions or what they forget to do to ensure that working equipment and the personal protective equipment are used as intended, and specific incidents reported immediately.

In addition to these responsibilities, the employees also have different rights, such as of proposal, of complaint, and a right to cease all work in the event of immediate significant danger. An employee must not be disadvantaged in any way should he or she make use of these rights.

In addition to these in-house procedures, which are regulated by the Federal Mining Act and the General Federal Mining Act, there is also the further instrument of the works safety staff and company doctors.

On the basis of a decree passed by the federal states, ("Mining Decree governing safety and company doctors," BVOASi) the operator must set up a works safety office and a works doctor's service to help carry out the duties of improving occupational safety, which includes averting work-related health risks and promoting accident prevention in the company.

These can be organized as external or in-house services. These services are unnecessary if the workforce is too small, or if the operator himself has participated in the relevant information, further training, and qualification courses.

The works safety office consists of qualified personnel for occupational safety and safety engineers and technicians, auxiliary personnel samplers and measuring assistants, and the concomitant facilities. Its duty is to advise the operator and "responsible persons" in managing the company; obtaining work equipment in particular, hazardous substances; to advise of deficiencies found (or anticipated) during regular inspections; to suggest improvements; and to collate employee suggestions.

In addition to this they are obligated to investigate the cause of work accidents and occupational illnesses, and be involved in briefings and maintenance of work safety facilities. These duties can also be carried out by "responsible persons." Within the confines of these special duties the qualified personnel for occupational safety are, however, completely autonomous. A prerequisite for this task is the relevant expertise; personnel are chosen only by special appointment. The qualified personnel for occupational safety are notably contact persons for the social-political advisory councils and pit Inspectors of the Mining Authorities.

The works medical office must have at least one company doctor as well as auxiliary personnel. They are obligated to advise the operator when planning, constructing, and maintaining production, social, and sanitary facilities. They are involved when obtaining working equipment, choosing and testing personal protective equipment, in questions of work physiology, psychology, ergonomics, and hygiene, and also carry out employee physical examinations.

In a works safety commission in companies with 21 employees and above, the operator works manager, representatives of the works committee, qualified personnel for work safety, company doctors, and safety officers must all be included. The commission must meet at least quarterly.

22 Evaluation

Germany cannot be described as a natural resources-deficient country. Even if the well-publicized hard coal mining is not cost effective and has experienced a large decline over recent decades, Germany is nevertheless the biggest lignite producer in the world: in potash and rock salt mining Germany is, respectively, the third and fourth largest producer. Even in a range of industrial minerals, German industry is among the leaders. We are self-sufficient, notably in rock and earthen mining, and in potash and rock salt mining, but nearly fully reliant upon imports for metals. We produce about a third of those energy raw materials we require.

The Federal Mining Act forms the basis for exploration, exploitation, and processing mineral raw materials. The ambit of this Act, however, includes only a selection of mineral deposits of specific importance. The vast majority of building raw materials in the stone and earthen sector, and concomitantly the highest proportion of production (in tonnage) of the exploited raw materials are subject to the substantially different decrees of the building, water, and immissions control laws. Only in the former GDR, where until 1996, on the basis of the transition regulations of the German Reunification Act, all mineral deposits mining excluded water subject to mining supervision.

It must be noted here that a unified legal settlement—planned before the enactment of the Federal Mining Act—for all natural resources excluding water would be an advantage for companies, the economy, and administration from today's standpoint, just as, for example, all waters are subject to the regulatory ambit of the Water Resources Act. This fragmentation is a consequence of the conflict of interests between the trade associations that existed when the Act was enacted, which today despite a change in interests, can no longer be remedied.

Due to the repeal of the §130 *BBergG* concerning cavity mining, the unified Mining Authority for setting up and operating underground cavities has been dropped. Consequently, tunnel and underground railway construction, for example, are subject to a partially fragmented jurisdiction for occupational safety and protection of the environment as well as for supervision and approval, which is now considered a disadvantage because the technical procedures and requirements are the same as those for mining.

The authorities responsible for mining supervision, despite unified federal law, are federal state authorities. Their structure is dictated not only by the scale of the mining, but also by federal, state, and political considerations and has developed very differently in the last 10 years. Federal states with little mining partially transferred the responsibilities of the relevant administration via treaties to neighboring federal states, nevertheless ensuring that the concerns of securing natural resources are represented by powerful bodies. Other federal states have, in contrast to this, disbanded their Mining Authorities as relevant administrations and transferred the duties to general authorities such as regional councils. Consequently the aspect of securing natural resources is no longer considered as important as other competing public interests.

In order to combine overall congeneric state facilities, the Geology Department was amalgamated with the Mining Authorities in several federal states. As subject-specific tasks vis-á-vis the internal administration of the Mining Authority are highly dominant, as wide a span as possible concentration of subject-specific tasks, if necessary including the Geological Offices, is significantly more efficient than pooling the tasks of the internal administration in general authorities.

23 Conclusion

A few powerful bodies with a broad remit also for water rights concessions, immission control, and waste have clear advantages over many small working units within the general administration. This speaks in favor of powerful independent Mining Authorities with the widest possible remit, if necessary including the Geological Office.

The legal responsibilities of the Mining Authorities include, in addition to the above-mentioned, securing natural resources occupational safety and health protection as well as the environment. These are implemented by the state via supervision and approval. Concentrating these responsibilities under one roof increases efficiency, is public-oriented, and advantageous for the execution of administrative procedures.

Adv Biochem Eng Biotechnol (2014) 142: 197–200
DOI: 10.1007/978-3-662-44474-0

Index

A

Acidithiobacillus ferrooxidans, 147, 151, 156, 160
Alcaligenes denitrificans, 147, 154
Altensalzwedel, 104
Anaerobic cultivation, 1
Anaerobic oxidation of methane (AOM), 111
Athabasca oil sands, 23
Authorities, 169

B

Bacillus, 8, 10, 30, 155
Benzoate, 154
Benzylsuccinate synthase (BSS), 131, 136
Bioaugmentation, 139
Biocracking, 34
Biodegradation, 6, 128
  in situ, 123
Biodepyritization, coal, 156
Bioreactor bay, 160
Bioremediation, 123, 128, 133
Bioremoval, sulfur, 154
Biosolubilization, 152
Biosurfactants, 30
  salinity/pH/temperature, 33
Brevibacterium sp., 147
BTEX (benzene-toluene-ethylbenzene-xylenes), 125
2,3-Butanediol, 32

C

Capillary number, 25
Capillary pressure, 27
Carbon capture and storage (CCS), 95
Carbon dioxide, 4, 22, 96
  disposal, 99
  leakage, 101
  storage, 100
Catalyzed reporter deposition FISH (CARD-FISH), 18
4-Chlorocatechol, 139
Coal, 10, 21, 99, 110, 147
  desulfurization, 147
  into natural gas, 22
  macerals, 148
  Shinn's model, 151
  stockpile, depyritized, 160
  sulfur, 148
Comminution bay, 158
Compound- specific stable isotope fractionation, 135
Compound-specific isotope analysis (CSIA), 135
Core samples, 9
Critical micelle concentration (CMC), 30
Cultures, 13

D

Dechlorination, anaerobic, 128
Deep biosphere, 95
Denaturing gradient gel electrophoresis (DGGE), 20
Dense nonaqueous phase liquids (DNAPLs), 132
Desulfovibrio, 10
Desulfurization, 147, 153
Diagenesis, 7
Dibenzo-*p*-dioxins, 133
Dibenzofurans, 133

Dibenzothiophene (DBT), 147, 149, 154
Dibenzothiophene-5-oxide, 154
Dichloroethene, 139
Dieldrin, 128
Dilution-to-extinction method, 10
Disulfides, 149
DNA extraction, 21
  probes, 17

E
Enhanced natural attenuation (ENA), 134, 138
Enhanced oil recovery (EOR), 1, 4
Enterobacter cloacae, 33
Environmental risk assessment (UVP) (Umweltverträglichkeitsprüfung), 169, 185
Ethyl *tert*-butyl ether (ETBE), 129
Extracellular polymeric substances (EPS), 152

F
Federal Mining Act (Bundesberggesetz-BBergG), 169
Fluorescence in situ hybridization (FISH), 18
FODO, 154
Framboids, 148

G
Gas storage, underground, 95
Gels, formation, 29
Geothermal energy, 95, 112
Geotoga, 6
German Reunification Act (1990), 169, 174
Glass slide test, 16
Greigite, 149
Growth media, 14

H
Halocatechols, 139
Heavy metals, 125
Hydrocarbons, chlorinated, 123, 125
  degraders, 8
  list of microbial degraders, 43–64
  reservoirs, 95
  residual, 25
Hydrofluorocarbons (HFC), 98
Hydrogen, 12, 108, 133, 160
  oxidation, 111
Hydrogen sulfide, 6
Hydrogenophaga sp., 104, 107

I
Identification, molecular biological methods, 17
Immision Control Act (BImSchG), 169, 179
Interfacial tension, 1, 26
Iron sulfides, coal, 149
Isolation techniques, 10, 14

K
Kerogen, 6

L
Laplace-Young equation, 27
Legal principles, 169
Leptospirillum ferrooxidans, 147, 160
Liability, 190
Lichenysin A, 33

M
Mackinawite, 149
Marcasite, 149
Metabolic pathway, 147
Metal sulfides, 151
Methane, 4, 24, 35, 101, 108, 135
Methanobacterium, 8
Methanococcus, 8
Methanoculleus, 8
Methanogenesis, 7, 24
Methanogenic hydrocarbon degradation, 1
Methanogens, 6
Methanosarcina barkeri, 35
Methyl *tert*-butyl ether (MTBE), 129
Microbial conversion, methane, 22
Microbial cracking, 35
Microbial enhanced oil recovery (MEOR), 1, 4
Microbial life, oilfields, 5
Microbial strains, oilfields, 39–42
Microbially influenced corrosion (MIC), 6, 18
Microorganisms, oil reservoirs, 8

Micropollutants, 123, 125
Mineral oil, 123
Mineralization, 128
Mines, abandoned, 180
surveyor, 191
Mining, 169
authorities, 176, 189
charts, 191
companies, responsibilities, 190
concessions, 169, 180
licenses, 180
supervision, 169, 187
Mining Employer's Liability Insurance Association, 190
Monitored natural attenuation (MNA), 134

N
Naphthalene, 23
Natural attenuation, monitored/enhanced, 123
Natural gas (methane), 4, 22, 24, 35, 101, 108, 135
Nucleic acid probes/stains, 18

O
Oil droplets, elasticity, 29
Oil in place (OIP), 1, 4
unrecoverable, 1
Oil production worldwide, 5
Oil recovery, 1
Oil swelling, 24
Oil viscosity, decrease, 35
Operating plans, 169, 183
Organic pollutants, microbial degradation, 123
Organic sulfur, 147, 153

P
Pachuca reactor, 147, 156
Peak oil, 1
Pendant drop method, 28
Perfluorocarbons (PFC), 98
Permits, 187
Persistent organic pollutants (POPs), 125
Pesticides, 123
Petri dish test, 16
Petrotoga, 6, 8
pH range, 7
Pharmaceuticals, 123
Phenols, 125, 132
Phospholipid-derived fatty acids (PLFA), 21
Phosphorus, 110
Phylogenetic tree, 19
Pilot plant, 147
Plugging, 36
Polycyclic aromatic hydrocarbons (PAHs), 125
Polymerase chain reactions (PCR), 18
Porespace alteration, 36
Porto Torres biodepyritization pilot plant, 159
Positive displacement sampler (PDS), 104
Pseudomonas sp., 147
Pyrite, 149
Pyritic sulfur, 147, 150
Pyrosequencing, 19
Pyrrhotite, 149

R
RAG-1, 31
Reclamation, 171
Regulatory jurisdiction, 188
Reject water disposal, 160
Renewable energy, 95
Rhamnolipids (glycolipids), 30, 33
Rights/duties, 189, 192
Rock dissolution, 36

S
Salzwedel-Peckensen, 106
Sampling techniques, oil reservoirs, 9
Siderite, 100
Smythite, 149
Spot oil prices, 5
Stable isotope probing (SIP), 20
Subsurface sampling, drill cores, 9
Sulcis coal, 160
Sulfides, 149
Sulfur, 147
organic, 153
pyritic, 150
Surface active compounds (SAC), 4
Surface mining, 173
Surface tension, 65
Surfactin, 30, 33
Syntrophy, 129

T
Thermoanaerobacter, 8
Thermococcus, 8
Thermotoga, 8
Thiols, 149

Thiopyrones, 149
Toluene dioxygenase, 129
Transformation, 128
Troilite, 149

U
Underground gas storage (UGS), 95, 108

V
Viscosin (lipopeptide), 30
Viscosity, 4, 25, 30, 34
  water, 34
Vocational training, 180

W
Waste management, 179
Water rights, 169, 179
Wettability, 34
Wilhelmy plate method, 26
Winogradsky solution, 21

X
Xylene, 132